LIFE SCIENCE VOCABULARY DEVELOPMENT

Crosswords and Wordsearch Puzzles

ALMA J. BROWN

Life Science Vocabulary Development

Crosswords and Wordsearch Puzzles

By

ALMA JOYCE BROWN

Graphics by Anthony R. Brown

TABLE OF CONTENT

Crossword Puzzles

Word Search Puzzles

SCIENCE

Across

2. recorded observations and measurements
4. Microscopic organism
7. Study of the relationship of living things with one another and their environment
8. Scientific test standard of comparison
9. Tested & accepted scientific theory
10. Logical explanation of events
11. Suggested solution to a problem

Down

1. Factor being tested in an experiment
3. Study of plants
5. Study of microorganisms
6. Study of animals

WORD LIST

microbiology	control	microorganism	botany
zoology	law	theory	variable
data	hypothesis	ecology	

SCIENTIFIC MEASUREMENTS

Across

3. amount of space an object takes up
6. measure of the force of attraction between objects due to gravity
8. basic unit of mass in metric system
9. Basic metric unit of length
10. metric temperature scale
11. force of attraction between objects
12. Unit of volume equal to 1/100
13. one thousandth

Down

1. mass per unit volume of a substance
2. exchange into an equivalent
4. basic metric unit of volume
5. one thousand
7. amount of matter in an object
9. way of finding size, quantity, or amount
10. 100th part of

WORD LIST

meter	milliliter	liter	volume	density
weight	kilo	milli	conversion	gravity
centi	measurement	kilogram	mass	Celsius

CHARACTERISTIC & NEEDS OF LIVING THINGS

Across

5. process of getting rid of waste
10. all chemical activities in an organism
11. reproduction requiring only one parent
12. an action or movement of an organism brought on by a stimulus
13. process by which living things take in oxygen and use it to produce energy
14. taking in food eating

Down

1. having a body temperature that changes with environmental changes
2. process by which living things produce offspring
3. a struggle to survive
4. organism's ability to keep inside body conditions constant
6. ability of an organism to maintain constant body temperature even though external condition change
7. signal to which an organism reacts
8. smallest part of a chemical element
9. food is broken down into simpler substances

WORD LIST

metabolism	ingestion	digestion	respiration	stimulus
response	competition	asexual	reproduction	excretion
atom	warm-blooded	coldblooded	homeostasis	

Chemistry of Living Things

WORD LIST

Matter
Atom
Proton
Neutron
Electron
Element
Compound
Molecule
Formula
Water
Sugar
Ferrum

Across

5. smallest particle of a compound having all the properties of that compound
10. Combination of chemical symbols
11. Positively charged particle in the nucleus of an atom
12. Tiny particle of matter with a nucleus containing protons and neutrons

Down

1. Two or more elements chemically combined
2. Neutral particle found in the nucleus of an atom
3. negatively charged particle of an atom
4. A compound made up of the elements carbon, hydrogen and oxygen
6. Pure substances that cannot be separated into simpler substances by ordinary chemical processes
7. The Latin word for iron
8. A compound composed of hydrogen and oxygen
9, anything that takes up space and has matter

Cell structure and function

Across

4. small round structure involved in digestive functions
5. reads the DNA's genetic information and guides the protein making process
10. all the protoplasm, or living material outside the nucleus of a cell
12. tiny cell structure
13. food making site in green plants
14. basic unit of structure and function of a living thing
15. entire living thing that carries out all the basic life functions
17. all chemical activities of an organism
18. substance used to build and repair cells
19. green substance needed for photosynthesis, found in green plants

20. thin membrane that separates the nucleus from the rest of the cell

Down

1. stores food, water and waste
2. structure that passes on traits to new cells
3. ability of a cell to maintain a stable internal environment
6. that control center the cell
7. the control center of the cell
8. long chain of sugar molecule manufactured by a cell that makes up the cell wall
9. stores the information needed to build and repair cells
11. Supplies most of the cell's energy
16. protein making structures of the cell

WORD LIST

DNA	Protein	Nucleus	Chloroplast
Mitochondria	organelle	lysosome	chlorophyll
RNA	cytoplasm	cell	homeostasis
vacuole	nuclear	organism	ribosome
chromosome	metabolism	cellulose	nucleolus

Cell Processes

Across

4. all chemical activities of an organism
5. process by which food molecules, oxygen water and other material enters and leaves a cell through the cell membrane
10. structure outside the nucleus in animal cells that play a part in cell division
11. fourth stage of mitosis resulting in the formation of two individual cells
12. second stage of mitosis
13. duplication and division of the nucleus and of the chromosomes during all reproduction

Down

1. fourth stage of mitosis resulting in the formation of two individual cells
2. special type of diffusion by which water passes into and out of the cell
3. process by which living things give rise to the same type of living things
6. all chemical activities of an organism
7. third stage of mitosis during which the chromosomes split apart
8. first stage of mitosis during which the nuclear membrane begins to disappear
9. process by which living organisms take oxygen and use it to produce energy

WORD LIST

Mitosis – telophase – osmosis – chromatin – metaphase – diffusion – interphase
prophase – metabolism – centriole – anaphase – reproduction – respiration

Classification of Living Things

Across

1. organisms made up of many cells
3. the science of classification
7. second largest classification grouping
8. the first person to develop a classification system for living things
9. classification grouping between order and genus
11. largest classification grouping
12. member of the monera kingdom
15. one-celled organisms
17. classification grouping between phylum and order

Down

2. organisms that can make its own food from simple substances
4. the group of living things according to similar characteristics
5. classification grouping between class and family
6. organisms that cannot make its own food
10. classified plants and animals according to similarities in structure
13. group of organisms that are able to interbreed and produce young
14. the first person to scientifically use the term species
16. the language that is most often used to organisms in the classification system

---

WORD LIST
binomial classification – taxonomy – species – phylum – class order – family –
multicellular – unicellular – autotroph – heterotroph – Aristotle Latin – moneran

Viruses and Bacteria

Across

4. bacteria that feeds on living things
6. process by which milk is heated to kill bacteria
7. tiny particle that contains hereditary material
8. jelly-like material inside the cell membrane
10. chemicals that destroy or weaken disease causing bacteria
11. whiplike structures that propel bacteria
12. rod shaped bacteria

13. nucleic acid that stores the information needed to build proteins and carries genetic information about an organism

Down

1. nucleic acid that reads genetic information
2. organism in which another organism lives
3. virus that infects bacteria
5. poisons produced by some bacteria
8. sphere shaped bacteria
9. unicellular organisms

WORD LIST

virus - host - DNA - RNA - bacteriophage - bacteria - bacilli - cocci - cytoplasm
flagella - parasites - antibiotics - pasteurization - toxins

Protozoan

Across

2. long thin whip-like structures that propel an organism
4. a type of sexual reproduction in which hereditary material is exchanged
7. a group of protozoan that cannot move by themselves
8. tiny reproductive cells that develop into protozoans
12. type of protozoan that lives in fresh water and moves by means of pseudopods
15. large, nucleus that controls reproduction in a paramecium
16. reddish structure in Euglena that is sensitive to light
17. the hard membrane that covers the other surface of the paramecium
18. long thin whip-like structures that propel an organism
19. protozoan that moves by means of pseudopods
20. a common sporozoan that causes malaria in human being

Down

1. store food. water and waste
3. small nucleus that controls reproduction in a paramecium
5. a sarcodine with a calcium carbonate stem
6. extension of the cytoplasm of a sacrodine that is used in moving and getting food
9. organism that feed on the other living organisms
10. an example of a flagellate that is both a heterotroph and an autotroph
11. funnel shaped structure in a paramecium through which food passes from the oral groove to the food vacuole
12. unicellular animal like organism; also means first animal
13. large nucleus that controls reproduction in a paramecium
14. protozoan that moves by means of pseudopods
21. sarcodine, a type of protozoan that lives in fresh water and moves by means of

---

WORD LIST

protozoan – conjugation – spore – sarcodine – ciliate – flagellate – sporozoan
pseudopod – amoeba – paramecium – pellicle – gullet – micronucleus – macronucleus
flagella – parasite – foramiciferan – chloroplast – eyespot – euglena – plasmodium

Algae and Fungi

WORD LIST

Nonvascular
holdfast
multicellular
chlorophyll
photosynthesis
diatom
flagella
dinoflagellate
bioluminescence
fungi
parasites
hyphae
club
cap
stalk
gill
sac
fermentation
yeasts
mushroom
molds
agar
bladders
alga
plankton

Down

1. Non vascular plant that contains chlorophyll
2. Energy releasing process in which sugars and starches are changed into alcohol and carbon dioxide.
3. A root like structure
5. Glow produced by fire algae
6. Plants that do not have transportation tubes t carry water and food
8. Organisms that feeds on other organisms
9. An example of club fungus
10. Type of fire algae that has two flagella
11. Green substance found in green plants
13. Structures that enable brown algae to float near the surface of the water are called air_____.
16. Most common and most attractive golden algae
17. Whip like structure that help propel an organism
18. A type of fungus that grows on some foods
21. Nonvascular plantlike organisms that have no chlorophyll

22. Obtain their energy through a process called fermentation
23. Small organisms that float or swim near the surface of the water

Across

4. Fungus that produces spores in a club-shaped structure
7. Substance on which scientist grow bacteria
12. Process by which organisms use energy from sunlight to make their own food
14. Fungus that produce spores in a sac-like structure
15. Umbrella-shaped part of a mushroom which is part of the mushroom's fruiting body
18. Many celled
19. Stem-like structure in a mushroom that supports the cap
20. Spore producing structure in a mushroom
22. Obtain their energy through a process called fermentation
23. Small organisms that float or swim near the surface of the water.

15

Vascular Plants

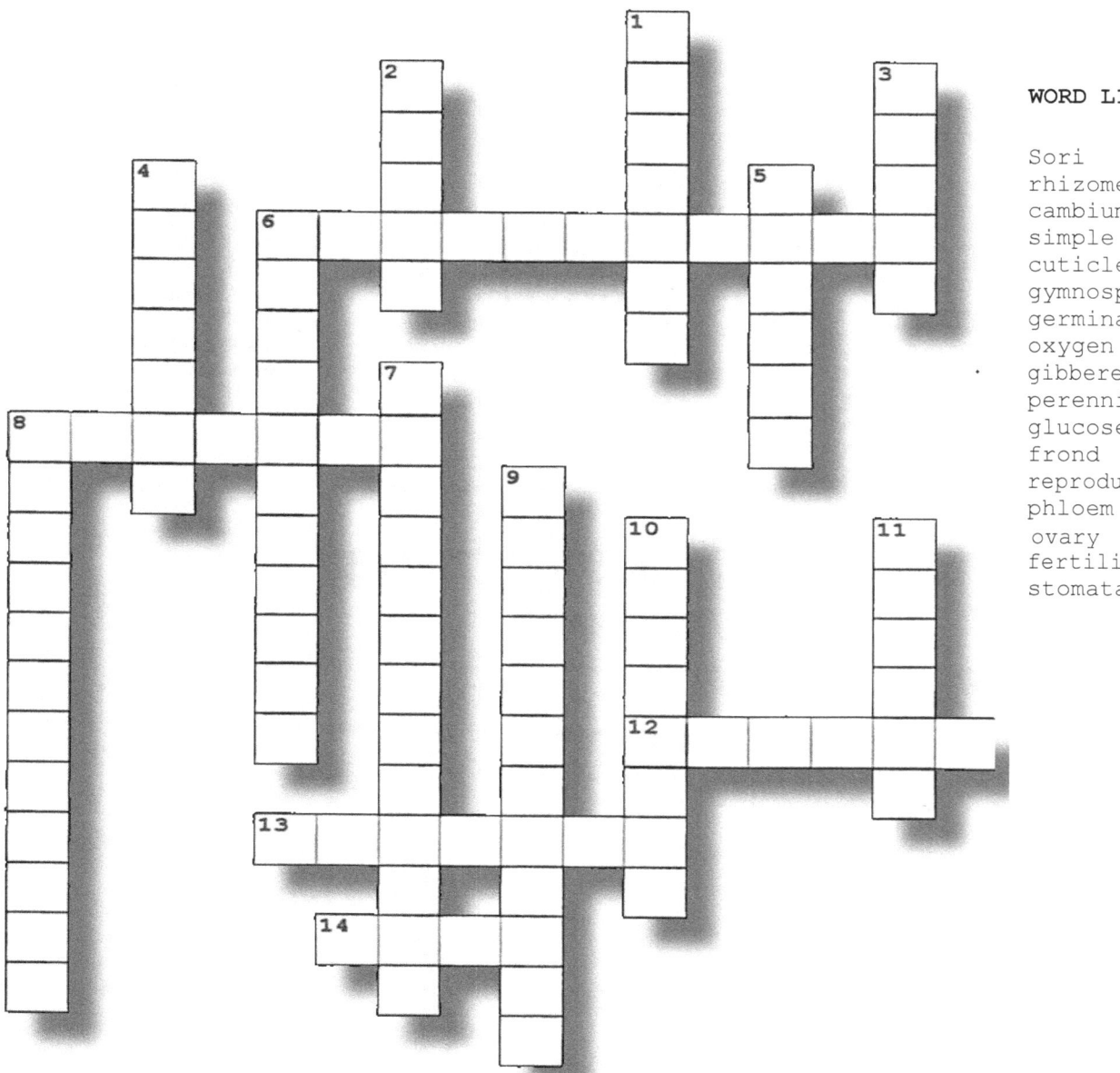

Down

1. Structure through which carbon dioxide enters the leaf during photosynthesis
2. Structure that produces eggs
3. The leaf of a fern
4. Produces new xylem and phloem cells in the stem
5. Leaves that have one blade
6. Cycads belong to this group of seed plants
7. The union of the sperm and egg
8. Plant hormones that cause stems to grow longer

9. Ferns produce spores during asexual_____
10. Fern structures that absorb water and minerals from the soil
11. Plant tissue that carries food substances down from the plant leaves

Across

6. A seed goes through this process of growth
8. Food made by a plant
12. The gas that is released during photosynthesis
13. The waxy covering
14. Small brown structure on the underside of fern fronds

Ferns

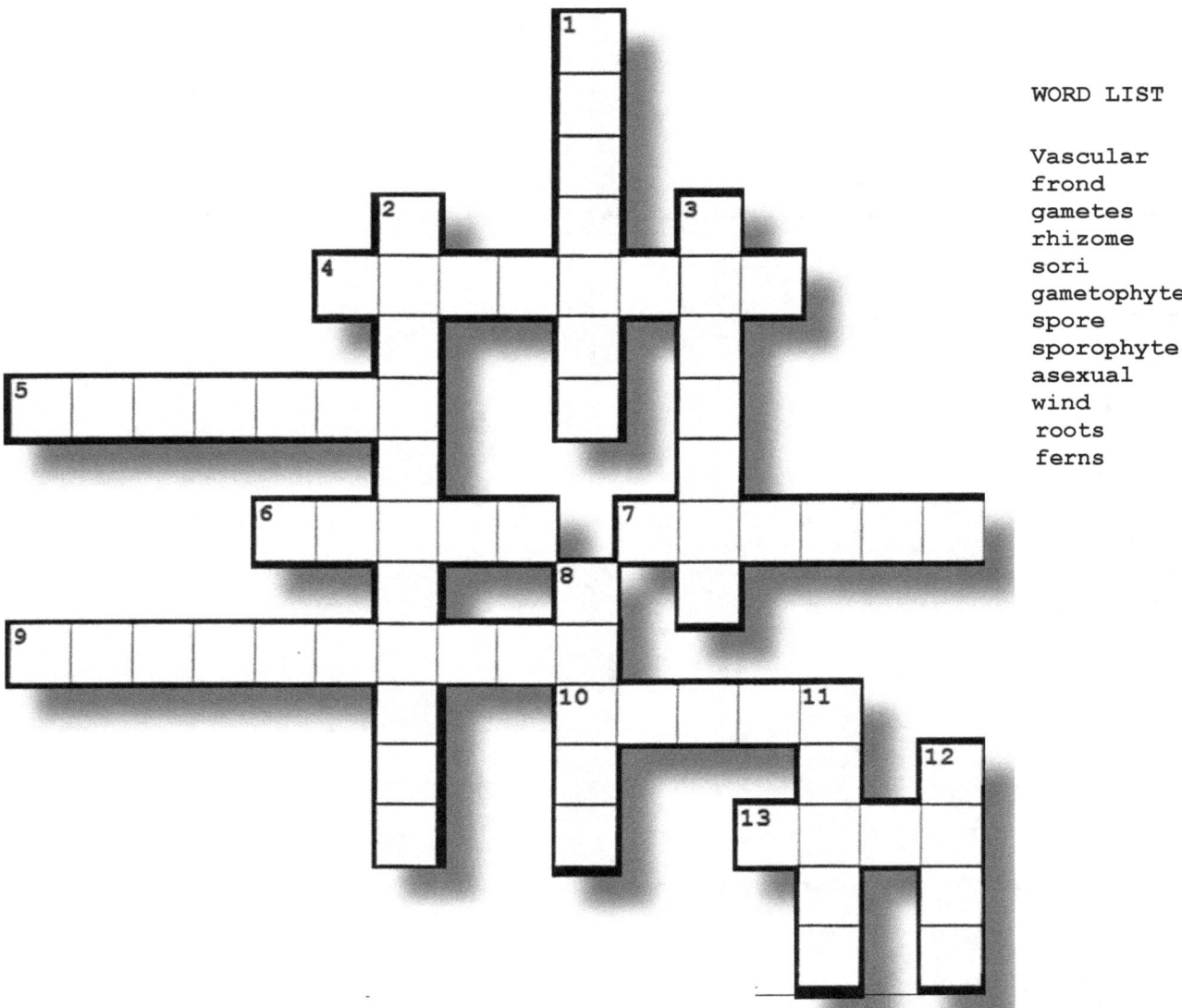

Down

1. Reproduction by which new fern plants develop from spores
2. Male and female sex cells are produced in this stage of a fern's life cycle
3. Male and female sex cells
8. Among the oldest plants on the earth
11. Tiny reproductive cells
12. Carries spores from place to place

Across

4. Plants contain transporting tubes that carry material throughout the plant
5. Underground stem of a fern
6. Leaf of a fern plant
7. Reproduction by which fern plants formed from the uniting of a sperm and an egg cell
9. Stage in a fern's life cycle during which each spore has the same type of hereditary material as the parents
10. Anchor the ferns to the ground
13. Small brown structures on the underside of a fern frond

Seed Plants

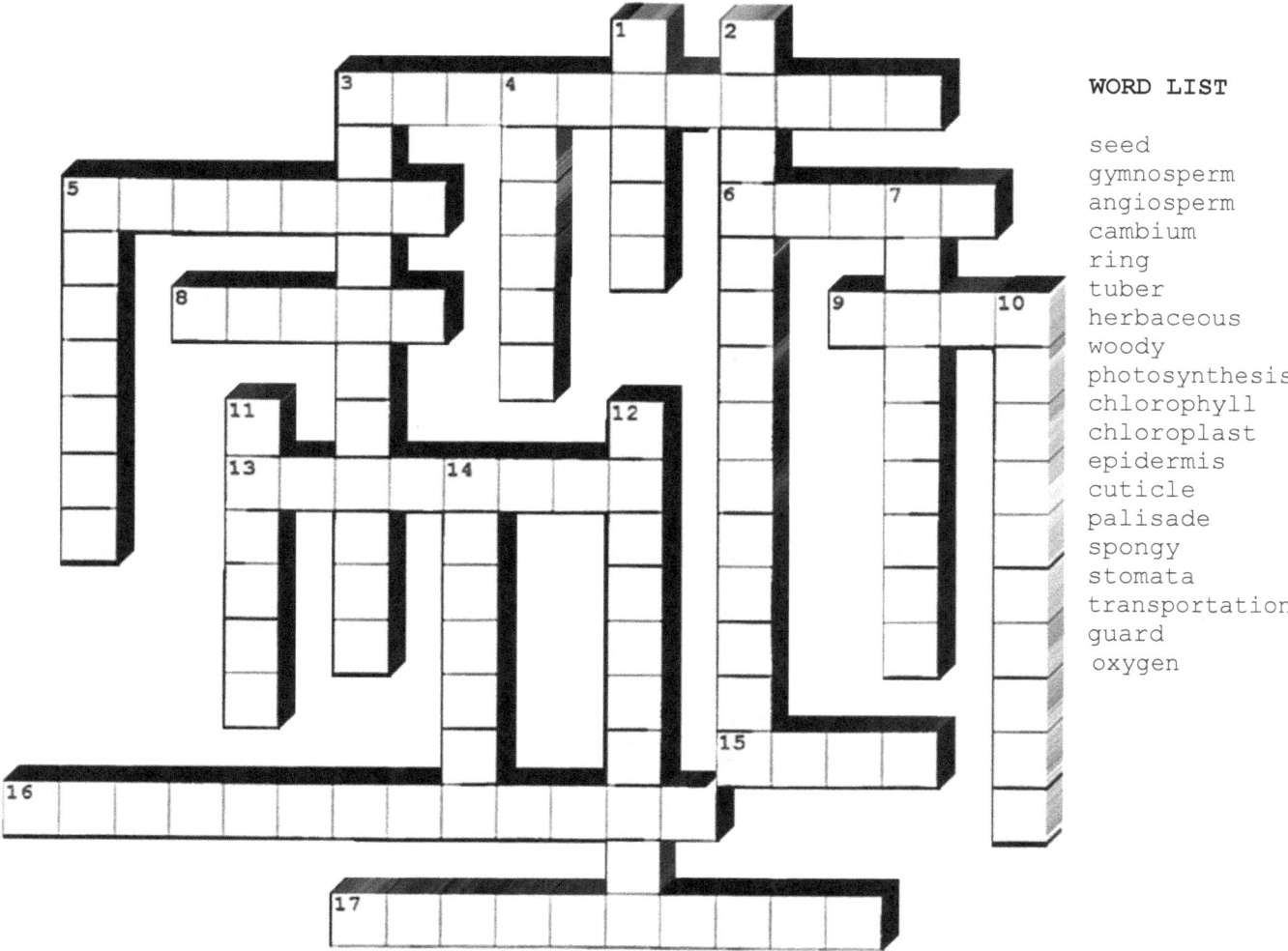

WORD LIST

seed
gymnosperm
angiosperm
cambium
ring
tuber
herbaceous
woody
photosynthesis
chlorophyll
chloroplast
epidermis
cuticle
palisade
spongy
stomata
transportation
guard
oxygen

Down

1. Rigid plant stem
2. Process by which organisms use energy
3. Large irregularly shaped structure that contains the green pigment chlorophyll: food making site in green plants
4. The gas that is released during photosynthesis
5. Growth tissue of the stem where xylem and phloem cells are produced
7. Outer protective layer of a leaf outermost layer of skin
10. Type of seed plant where seeds are not covered by a protective wall
11. Irregularly shaped food making cells of the lower mesophyll of a leaf is called the_____layer
12. soft, green plant stem
14. Opening in the lower surface of the epidermis

Across

3. Green substance needed for photosynthesis found in green plant cells
5. The waxy covering of a leaf
6. Underground stem of a plant
8. Sausage-shaped cell that regulates the opening and closing of stomata
9. One year's growth of xylem cells is called an animal_____
13. Long cylindrical food-making cells
15, structure from which a plant grows; contains a young plant, stored food and a seed coat.
16. The process by which excess water is released through the leaves
17. Seed plant whose seeds are covered by a protective wall

Gymnosperms and Angiosperms

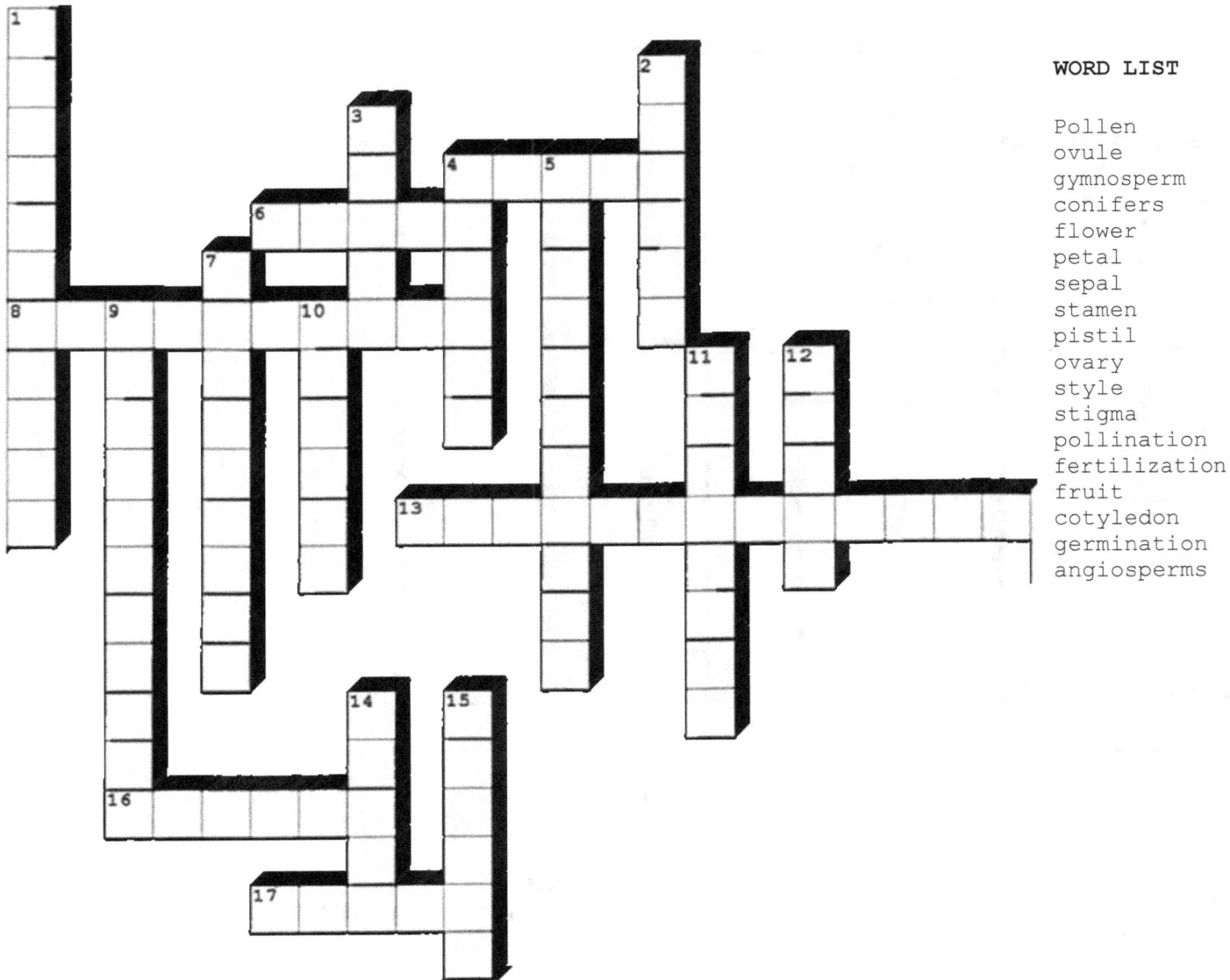

Down

1. Early growth stage or the sprouting of a young plant.
2. Contains the male sex cells of a seed plant
3. Structure that contains the female sex cells of a seed plant
4. Male reproductive organ of a flower
5. Transfer of pollen from male part to the female part of a flower
7. Leaf-like structure of a young plant that store food
9. Seed plant that produces uncovered seeds
10. Female reproductive organ of a flower, part of a flower that produces seeds
11. One type of gymnosperm
12. Colorful leaf-like structure that surrounds the make and

Across

4. Leaf-like structure enclosing a flower when it is still a bud
6. Ripened ovary of an angiosperm
8. Seed plants that produce covered seeds
13. Joining of the egg and the sperm
16. Sticky structure at the top pf the pistil
17. Slender tube that connects the ovary to the stigma

Invertebrates

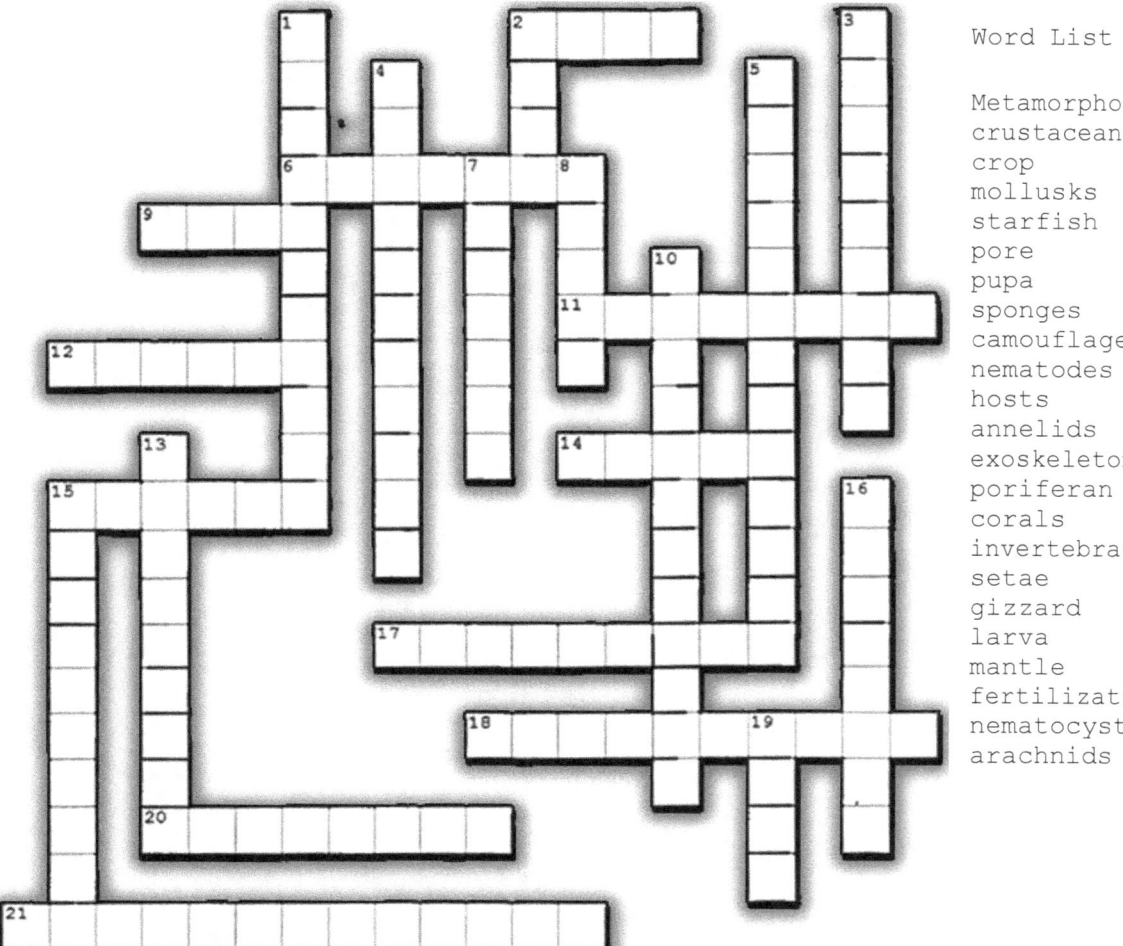

Word List

Metamorphosis
crustaceans
crop
mollusks
starfish
pore
pupa
sponges
camouflage
nematodes
hosts
annelids
exoskeleton
poriferan
corals
invertebrate
setae
gizzard
larva
mantle
fertilization
nematocyst
arachnids

Down
1 Crabs, lobsters, crayfish, and shrimp are included in this group of anthropods
2 opening of the outer surface of an animal through which materials enter and leave
3 Flatworms are classified as _____.
4 rigid, outer covering of an organism
5 joining of the egg and sperm
7 structure in an earthworm that grinds up food
8 bristles on the segment of an earthworm that help it pull itself along the ground
10 animal without a backbone
13 Spiders, scorpions, ticks, and mites are included in this group of anthropods.
15 hiding from enemies by blending in with the surroundings
16 invertebrates with a soft, fleshy body that is often covered by a hard shell
19 saclike organ that stores food in an earthworm

Across
2 stage that follows the lava stage in an insect's life
6 these poriferans grow attached to one spot in the ocean floor and remain there unless a strong wave of current washes them somewhere else
9 organism in which another organism lives
11 have segmented bodies and live in soil, salt water, or fresh water
12 part of a mollusk that produces material that makes up the hard shell
14 stage of insect that develops from an egg
15 these coelenterates live together
17 member of the phylum poriferan
18 stinging cell that is found in the mouth of a coelenterate
20 the best known echinoderm
21 change in appearance due to development

More about Invertebrates

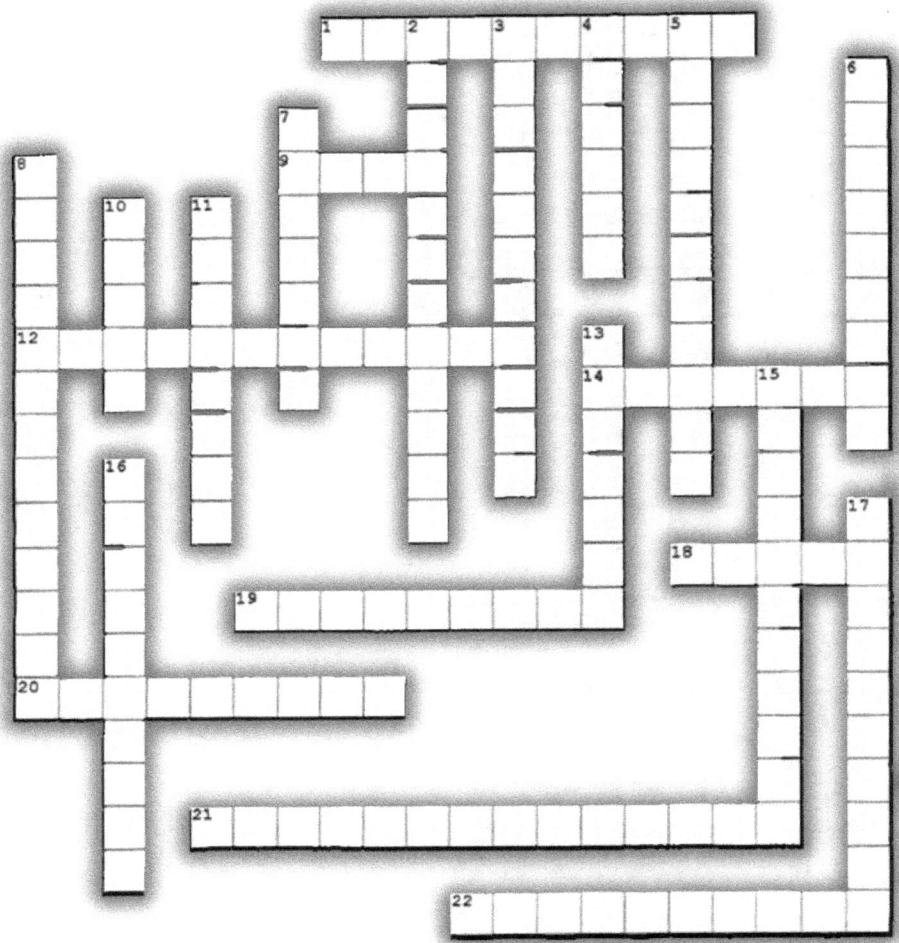

Down

2 ability of an organism to regrow
   lost parts
3 rigid, outer covering of an organism
4 filelike structure in the mouth of a
   univalve used to scrape food from
   an object
5 a disease caused by eating pork
   that contains roundworms
6. Swimmers try to avoid this coelenterate
   because it can deliver a painful poison
   through its stinging cells
7 These poriferans grow attached to one
   spot in the ocean floor
8 joining of the egg and sperm
10 stage of an insect that develops from
    an egg
11 echinoderm that eats the soft body
    parts of living coral
13 structure in an earthworm that
    grinds up food
15 Crab, lobsters, crayfish, and shrimp
    are classified as _____.
16 the special scents produced by insects
    to attract mates
17 These meat eating anthropods have
    one pair of legs in a segment

Across

1 animal with a backbone
9 opening of the outer surface of an
   animal through which materials enter
   and leaves
12 animal without a backbone
14 have 3 body parts, 3 pairs of legs,
    and an open circulatory system
18 bristles on the segment of an
    earthworm that help it pull itself
    along the ground
19 invertebrate that has jointed legs and
    an exoskeleton
20 These roundworms resemble spaghetti
21 have flat bodies and live in ponds
    and streams
22 These plant eating anthropods have
    2 pairs of legs in each segment

Coldblooded Vertebrates

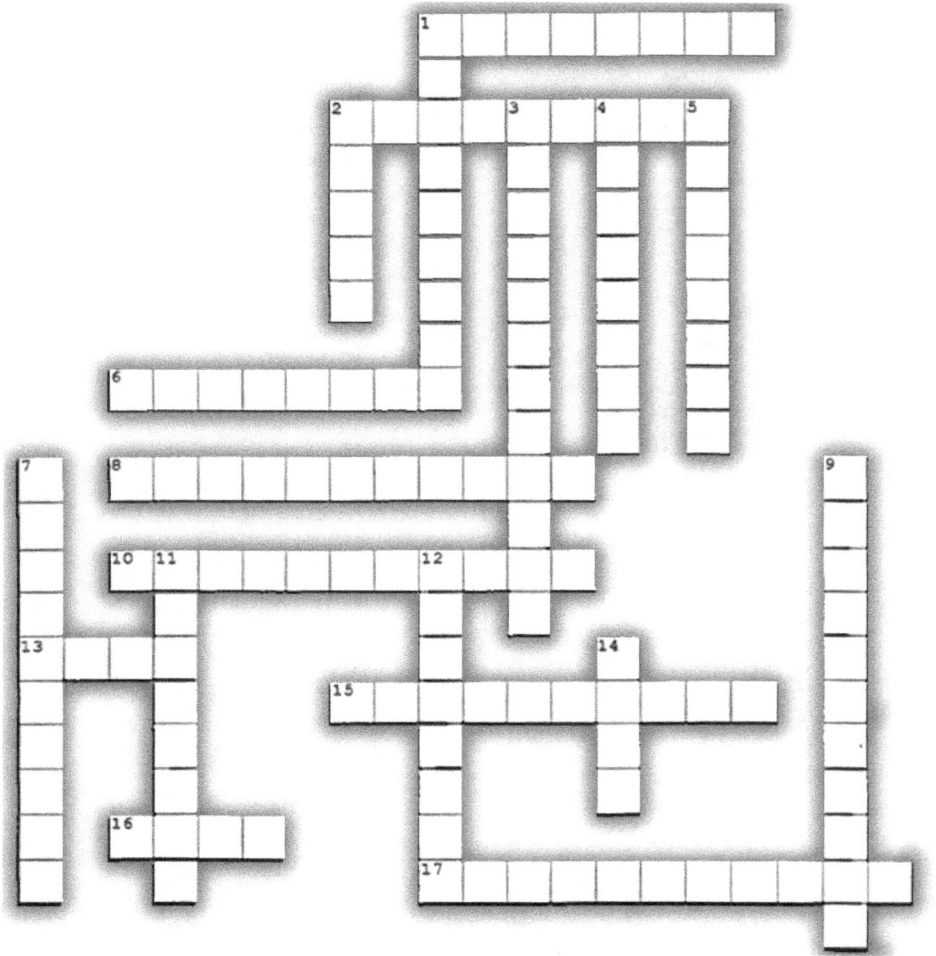

Down

1 This flexible tissue gives support and shape to body parts.
2 Snakes use special glands to produce this poison.
3 internal skeleton
4 a class of coldblooded chordates that have dry, scaly skin and lay eggs on land
5 fertilization that takes place outside the female
7 amphibian with a tail
9 having a constant body temperature
11 fertilization that occurs inside the body of a female
12 Frogs and toad eggs hatch into _____.
14 the vertebrates that are best adapted to life underwater

Across

1 phylum of vertebrates
2 bones that make up a vertebrate's backbone
6 part of a vertebrate's endoskeleton that helps protect the nerves of the spinal cord
8 having a body temperature that can change somewhat with changes in the temperature of the environment
10 All body activities slow down during this winter sleep.
13 periodically shed one's skin
15 spend part of their lives in water and part on land
16 special tooth in snakes used to inject venom into their prey
17 sac filled with air that enables bony fish to rise or sink in water

Warmblooded Vertebrates

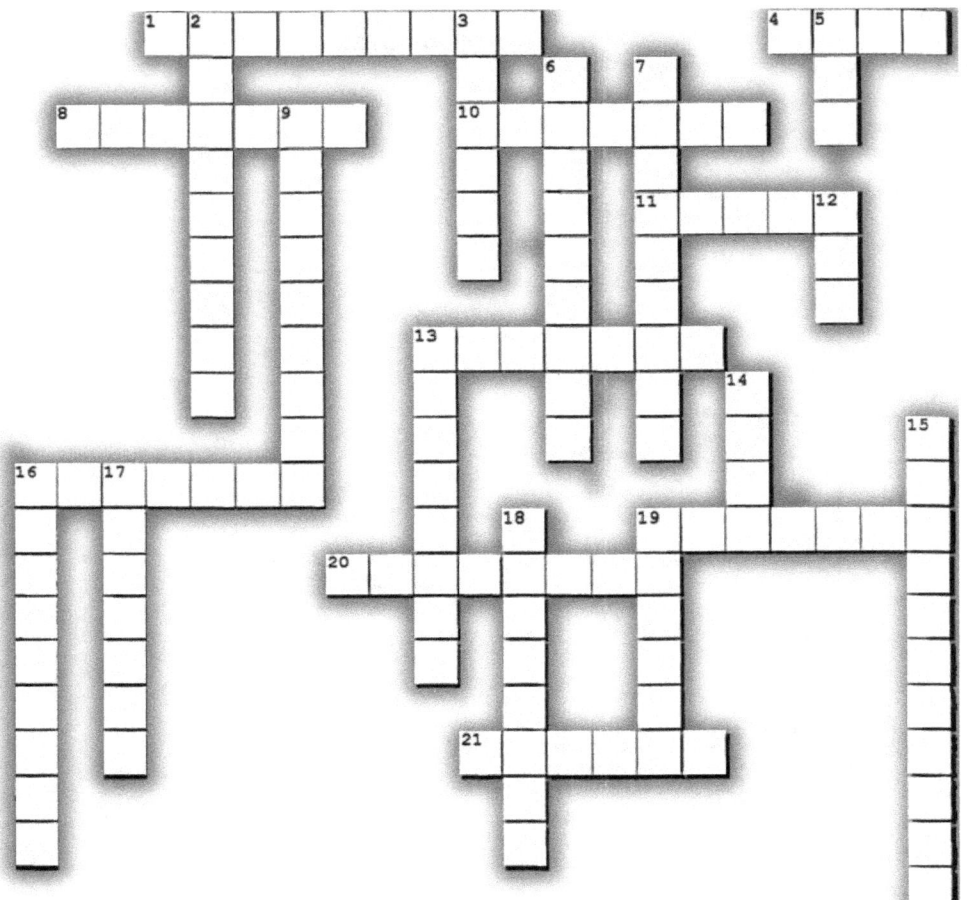

Down

2 toothless mammal
3 rodent-like mammal
5-12 hollow structure connected to a
    bird's lungs that increases the
    amount of air to the lungs
    (2 words)
6 pouched animal
7 organism that eats only plants
9 area where an animal lives
13 warm eggs by sitting on them until
    they hatch
14 short, fuzzy type of feathers used
    for insulation
15 having a constant body temperature
16 egg laying mammal
17 move to a new environment during
    the course of a year
18 insect eating mammals
19 This sharp pointed tooth is used for
    tearing and shredding meat.

Across

1 flesh eating animal
4 the only flying mammals
8 Humans, monkeys, apes belong to this
    order of mammals.
10 gnawing mammals that use chisel-like
    incisors for chewing
11 warmblooded vertebrates that have wings
    and are covered with feathers
13 front tooth used for biting
16 gland structure in a female mammal that
    produces milk
19 These large feathers, used for flight, are
    found on a bird's wings and on most of
    its body.
20 developing mammals receive food and
    oxygen while in the mother through this
    structure
21 water dwelling mammals

The Skeletal & Muscular Systems

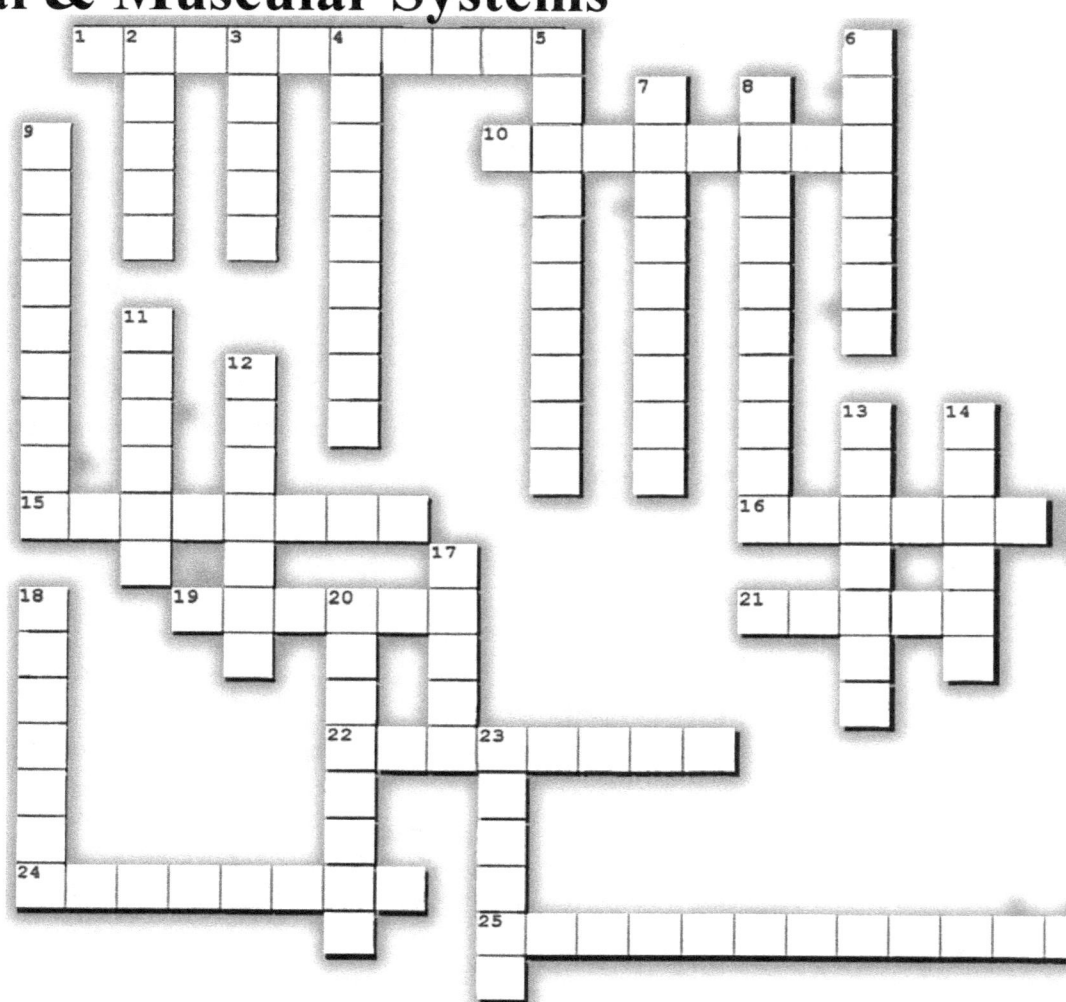

WORD LIST

muscle
connective
nerve
epithelial
organ system
ligament
vertebra
cartilage
ossification
periosteum
Haversian
marrow
joint
calcium
skeletal
tendon
smooth
cardiac
mandible
phalanges
patella
humerus
femur
clavicle

Down

2-11. The 4^{the} level of organization in an organism, system in which a group of organs work together to perform certain functions

3. Tissue that carries messages back and forth between the brain and spinal cord and to every part of the body.

4. Flexible tissue that gives support and shape to body parts

5. Tissue that forms a protective surface for the body and line the cavities of other body parts

6. knee cap

7. Canal of passageway running through the thick bone containing blood vessels and nerves

8. The tough membrane that surrounds the shaft of the bone

9. Toe bones

12. Upper arm bones

13. Muscles found in the heart

14. Muscles responsible for involuntary movement

17. Thigh bone

18. Necessary for bone growth and keeping bone strong

20. Collar bone

23. Connective tissue that connects muscle to bone

Across

1. Tissue that provides support for the body and unites its parts

10. Stringy connective tissue that holds the bones together

15. Muscle that is attached to the bone and moves the skeleton

16. The soft material found inside the cavity of a bone

19. Tissue that has the ability to contract and make the body move

21. The place where two bones meet

22. Bone that make up the vertebrate's backbone

24. Jaw bone

25. The changing of cartilage into bon

The Digestive System

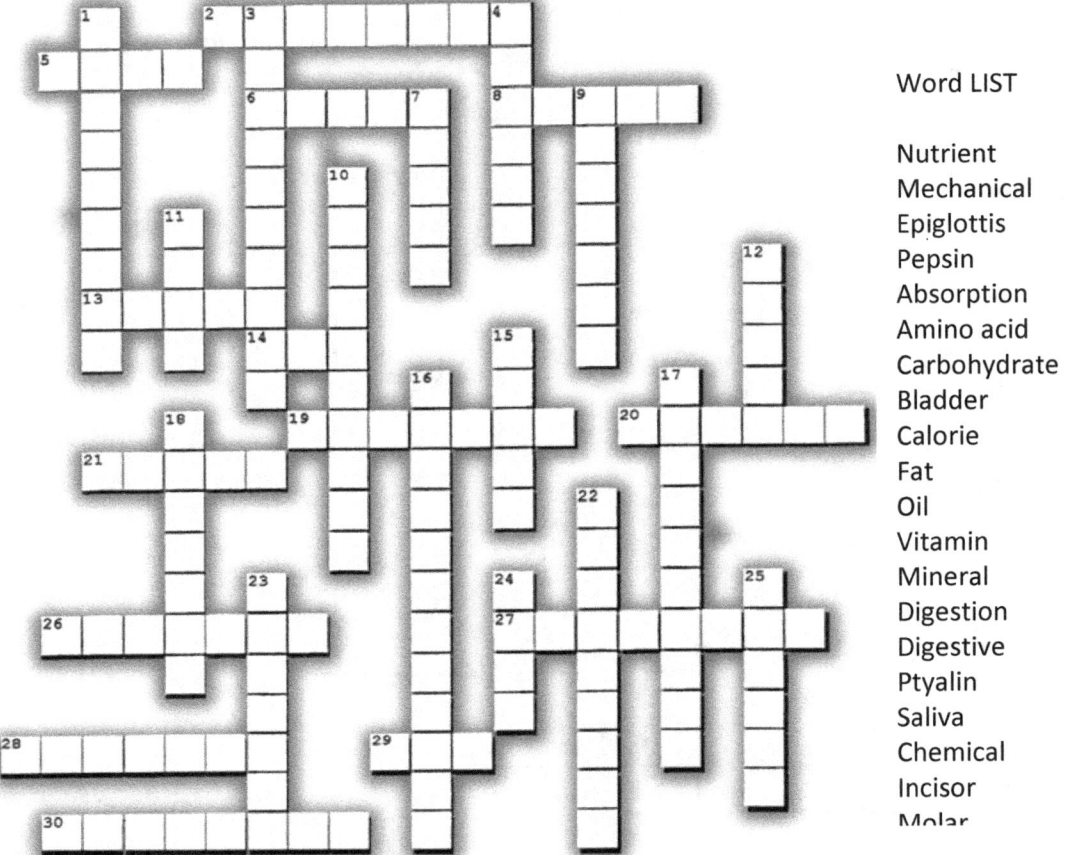

Word LIST

Nutrient
Mechanical
Epiglottis   Small
Pepsin   Liver
Absorption   Pancreas
Amino acid   Bile
Carbohydrate   Gall
Bladder
Calorie   Villi
Fat   Large
Oil   Feces
Vitamin
Mineral
Digestion
Digestive
Ptyalin
Saliva
Chemical
Incisor
Molar

Down

1 system in which food is broken down into simpler substances for use by the body
3 the process during which nutrients are taken into the bloodstream
4 liquid produced by the salivary glands that moistens foods and contains an enzyme that breaks down starches
7 intestine in which water is absorbed and undigested food stored
9 nutrient that helps growth and normal body functioning
10 small flap if tissue that closes over the windpipe
11-18 organ that stores bile
12 solid waste that is eliminated from the body
15 -24 building block of protein
16 energy-rich substance found in foods such as vegetables, cereal grains, and bread
17 this type of digestion involves the physical action of breaking food down into smaller pieces
22 process by which food is broken down into simpler substances
23 the energy value of food is measured in
25 sharp pointed tooth used for tearing and shredding meat

Across

2 organ that produces pancreatic juice and insulin
5 substance produced by the liver that aids in digestion
6 intestine in which most digestion takes place
8 organ that produces bile
13 Food is absorbed in the bloodstream through these hairlike projections in the small intestine.
14 energy rich-substance
19 enzyme in saliva that breaks down some starches into sugar
20 This enzyme, produced by the stomach, digests protein.
21 back tooth that grinds and crushes
26 This simple substance, found in nature, helps maintain the normal functioning of the body.
27 This type of digestion involves breaking down food by enzymes.
28 front tooth used for biting.
29 the body stores excess energy in this form
30 usable portion of food

The Circulatory System

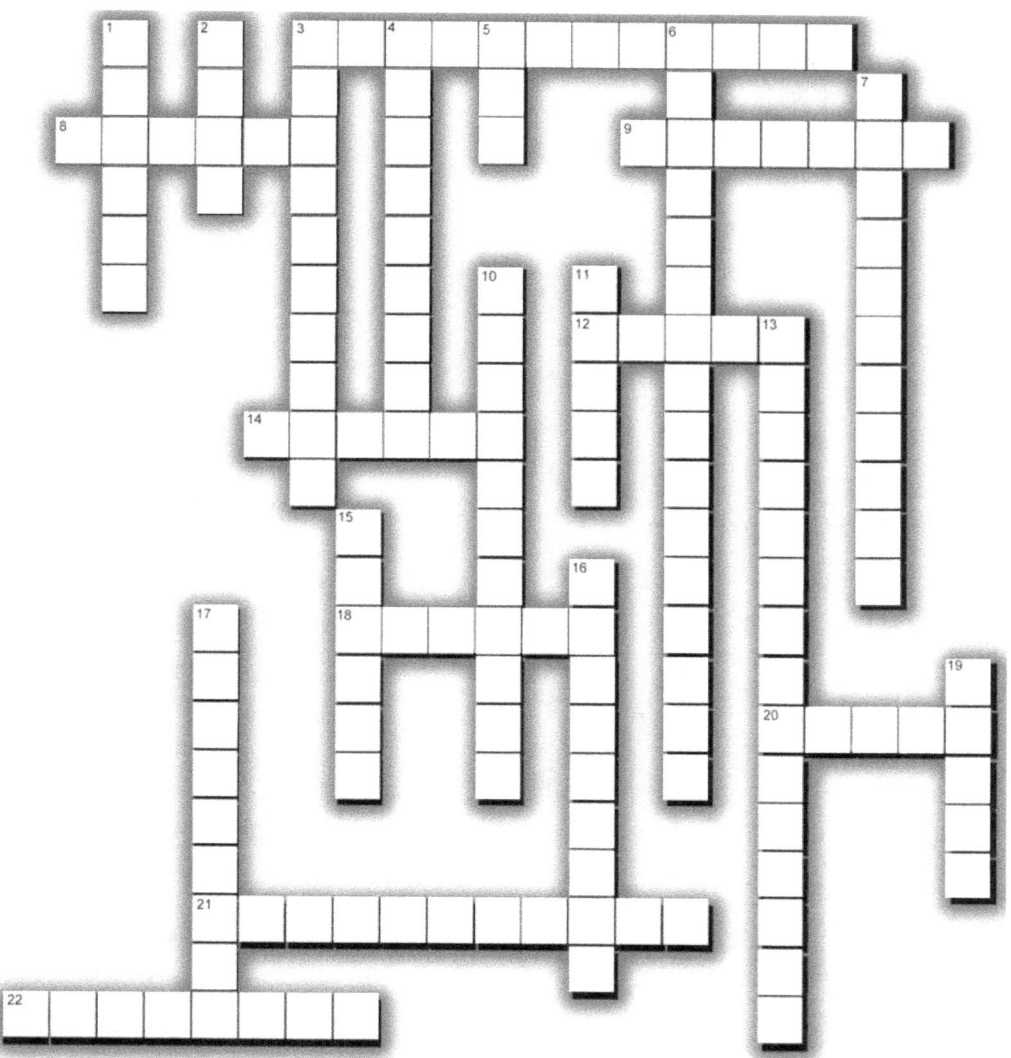

Down

1. Blood vessel that carries blood away from the heart.
2. Blood vessel that Carries blood to the heart
3. Iron containing protein in red blood cells
4. Special tissue in the heart that controls the case at which the heart beats
5. Blood cell that Carries oxygen throughout the body
6. instrument for measuring blood pressure
7. body system that delivers food and oxygen to body cell and Carries carbon dioxide and waste products away from body cells
10. The process of transferring blood from one body to another
11. Small flap of tissue between the upper and lower chambers of the heart
13. Thickening of the inner lining of the arteries
15. Thick wall of tissue that separates the heart into right and left sides
16. Tiny thin walled blood vessel
17. Lower chamber of the heart of the
19. Blood cells that act as a defense system against disease

Across

3. High blood pressure
8. Upper heart chamber
9. disorder or illness that has lingering or lasting
12. Large as blood vessel in the body
14. substance that traps blood cells and plasma, forming a scab
18. yellowish liquid portion of blood
20. Plasma that leaks out of the blood and surrounds and bathes body cells
21. fatty substance found in animals fats, meats and dairy products
22. Blood cell fragment that aid in blood clotting

The Respiratory and Excretory Systems

WORD LIST

liver
Diaphragm
alveolus
bronchus
capsule
dermis
epidermis
water
epiglottis
excretion
excretory system
exhale
inhale
kidney
larynx
lung
nephron
nostril
respiration
respiratory system
skin
trachea
urea
urethra

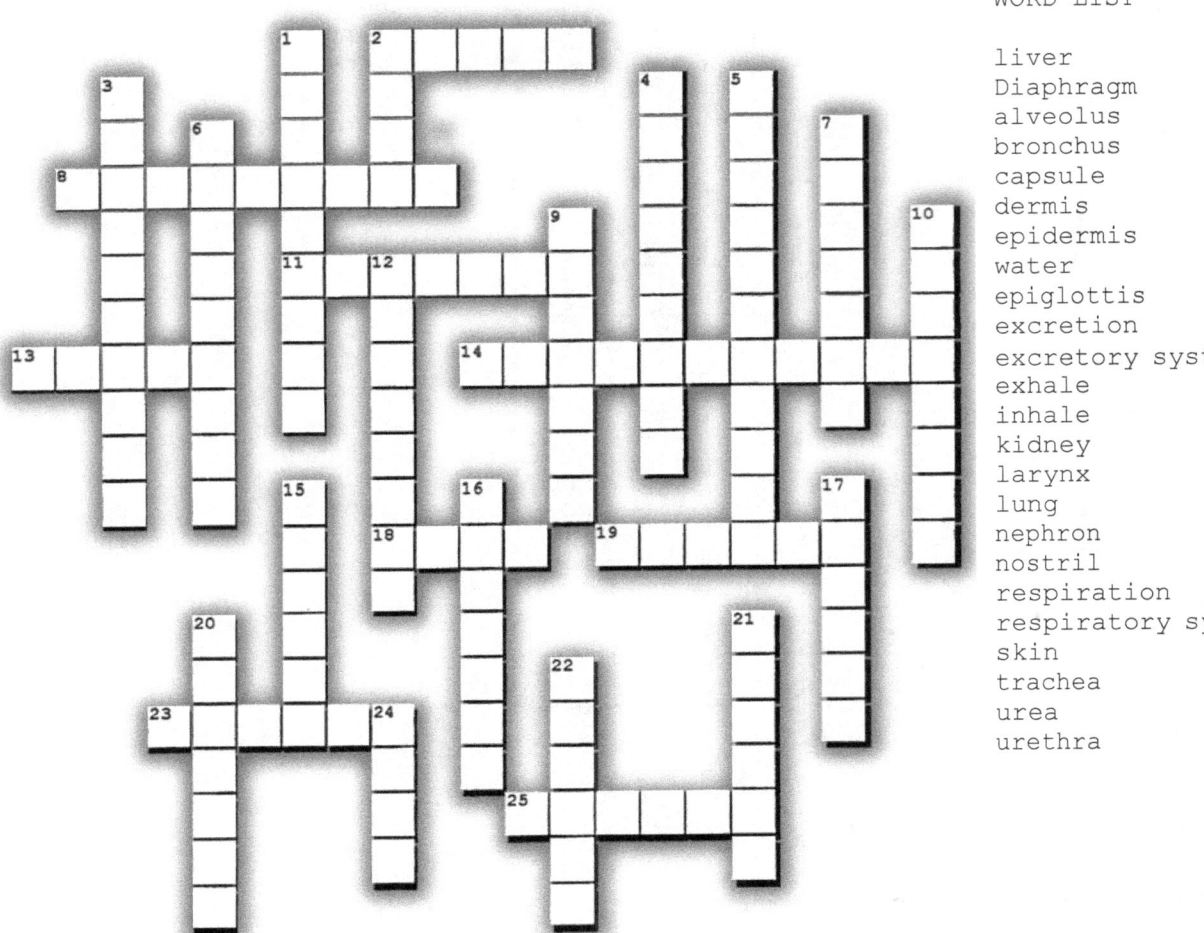

Down

1-15 system that removes body wastes
   (two words)
2 main respiratory system
3 flap of tissue that closes the windpipe
4 getting rid of waste materials
5-21 gets oxygen into the body and
   removes carbon dioxide and water
   from the body (2 words)
6 outer protective layer of a leaf or outermost layer of skin
7 opening in the nose
9 cup-shaped part of the nephron
10 tube that branches off from the trachea
12 the air cells in the lungs
16 microscopic chemical filtering factory
   in the kidney
17 breathe out
20 tube through which urine passes out
   of the body
22 most import excretory organ
24 outer covering of the body

Across

2 organ that produces bile
8 muscle at the bottom of the chest that
   aids in breathing
11 windpipe, carries air to the lungs
13 about 60% of your body weight
14 living organisms take in oxygen and
   use it to produce energy
18 nitrogen waste formed in the liver
19 voice box
23 inner layer of the skin
25 breathe in

The Nervous System

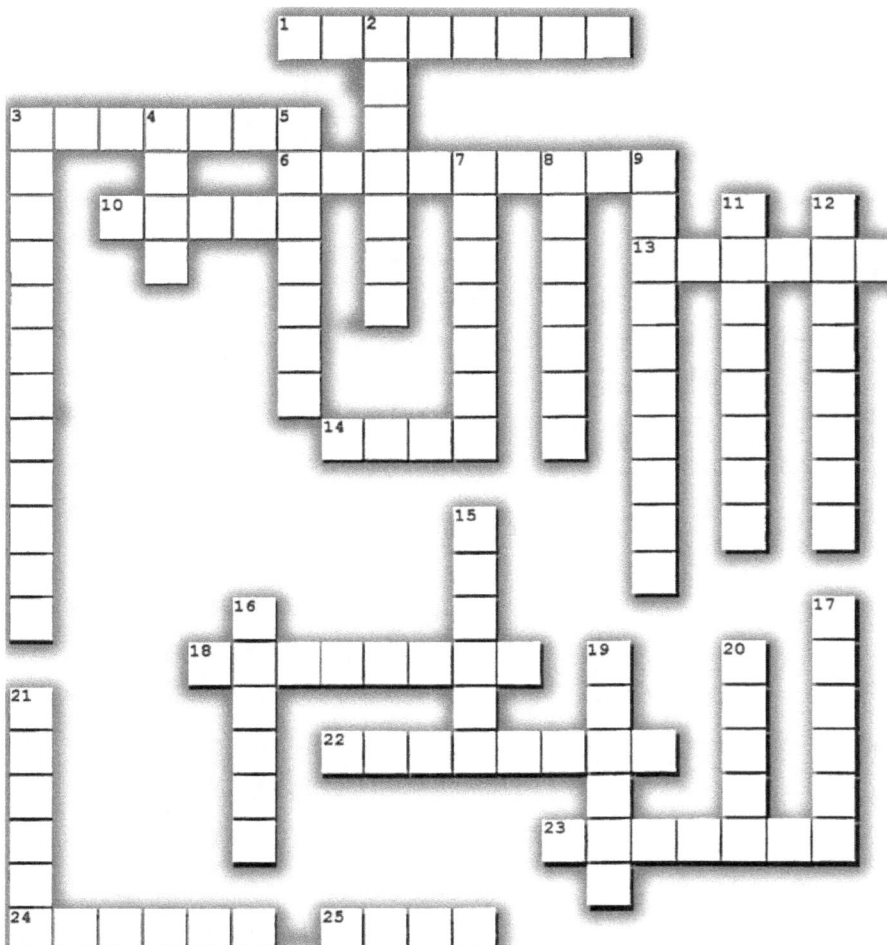

Down

2 This part of the nervous system is made up of the brain and spinal cord

3-21 curved tube in the inner ear that is responsible for balance (2 words)

4 fiber that carries messages away from a body cell

5 membrane in the ear that vibrates when struck by sound waves

7-15 system that controls all of the activities of the body (2 words)

8 part of the brain located at the base of the brain stem that controls involuntary body processes

9 part of the brain that controls balance and posture

11 part of the body that carries out the instructions of the nervous system

12 part of the brain that controls the senses, thought, and conscious activities

16 inner eye layer on which an image is focused

17 transparent protective covering of the eye

19 nerve cell

20 circular opening at the center of the iris

Across

1 part of the nervous system that respond to stimuli

3 tiny gap between an axon and a dendrite

6 division of the peripheral nervous system that controls all involuntary body processes

10 type of neuron that carries messages from the central nervous system to effectors

13 automatic reaction to a stimulus

14 part of the eye that focuses the light ray coming into the eye

18 fiber that carries messages from a neuron toward the cell body

22 signal to which an organism reacts

23 spiraling tube in the inner ear from which nerve impulses are carried to the brain

24-25 part of the nervous system that connects the brain to the rest of the nervous system

The Endocrine System

endocrine system
hypothalamus
parathyroid
ovary
insulin
testosterone
hormone
pituitary
thyroid
adrenals
pancreas
testes
estrogen
thyroxine
gonadotropic
endocrine gland

Across

7-9 body system that produces chemicals that control the body (2 words}

12 hollow structure that contains the egg cells of a flower. female sex gland

15 chemical messenger that travels through the blood

17 this hormone produced by the pituitary glands, stimulates development of male and female sex organs

Down

1 organ that produces pancreatic juice and insulin

2 hormone responsible for the growth of facial and body hair, broadening of the shoulders, and deepening of the voice in males

3 hormone produced by the pancreas

4 endocrine gland on top of each kidney that produces the hormone adrenaline

5 endocrine gland, located below the hypothalamus, which produces hormones that regulates growth and other body functions

7-16 gland that releases hormones directly into the blood stream (2 words}

8 hormone produced by the ovaries

10 endocrine gland producing a hormone that controls the level of calcium in the blood

11 endocrine gland at the base of the brain that controls body temperature, water balance, appetite, and sleep

13 is produced by the thyroid

14 endocrine gland that produces a hormone that controls metabolism

Diseases, Defense and Disorders

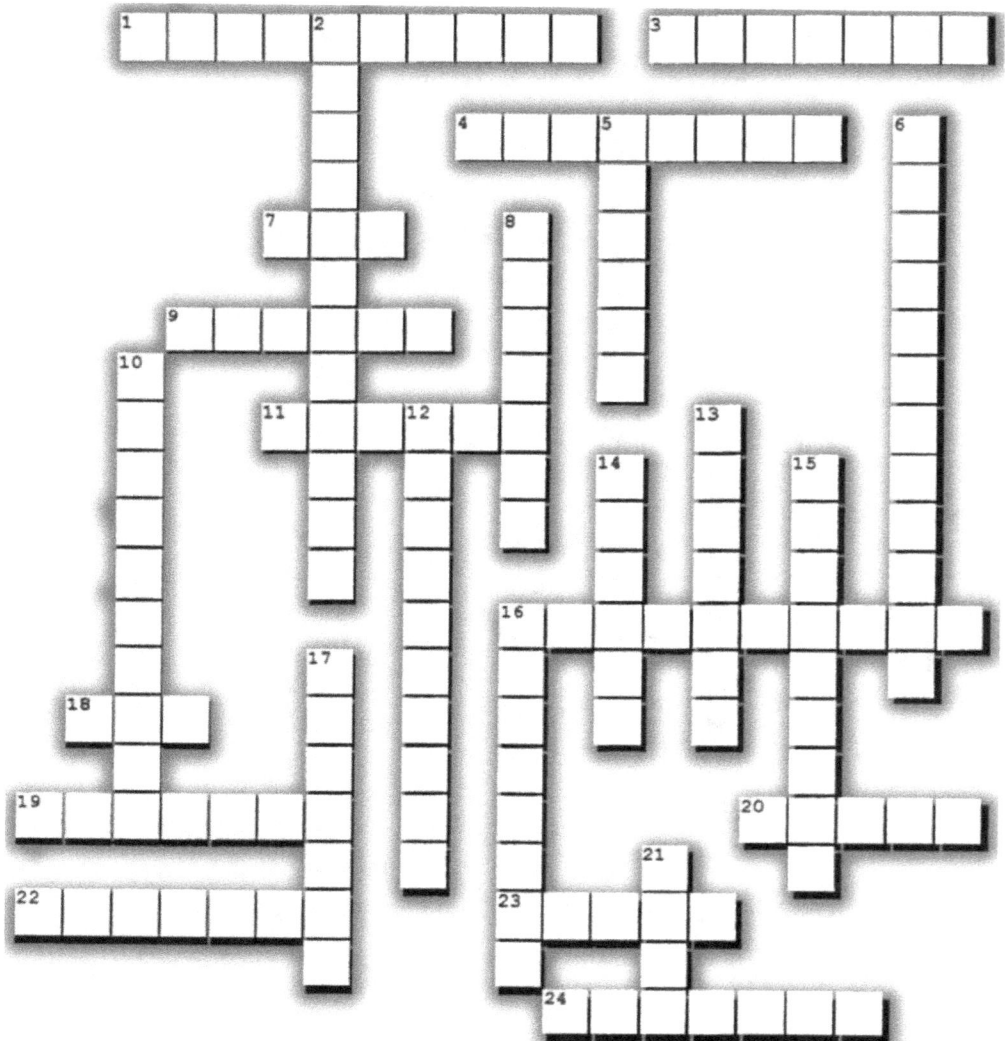

host
antibiotic
STD
immunity
interferon
acquired immunity
cancer
carcinogen
infection
toxin
tetanus
inflammation
antibody
tumor
allergy
infectious disease
communicable disease
immune system
pus
antigen
vaccine
benign

Across

1-3 disease that is transmitted among people by harmful organisms such as viruses and bacteria: communicable disease

4 protein produced by certain kinds of white blood cells in response to an invasion by a particular organism or substance

7 thick yellowish-white substance in an infected area made up of dead bacteria, destroyed body cells. and dead white blood cells

9 tumors that are not cancerous

11 abnormal and uncontrolled cell reproduction

16 substance produced by a body cell when invaded by a virus

18 infectious disease passed from person to person through sexual contact

19 substance that increases immunity

'20 poison

22 dangerous bacterial disease spread through a puncture wound

23 swelling of tissue that develops separately from the tissue surrounding it

24 invading organism or substance

Down

2-17 infectious disease (2 words)

5-14 body's defense system against disease (2 words)

6 body's response to an attack by disease causing organisms

8-16 immunity that develops during a person's lifetime (2 words)

10 chemical that destroys or weakens disease causing bacteria

12 cancer causing substance

13 reaction that occurs when the body is especially sensitive to certain substances called allergens

15 state caused when the body is invaded by disease-causing organisms

16 body's ability to fight off disease without becoming sick

21 organism in which another organism lives

Drugs, Alcohol and Tobacco

Word List

heroin
prescription drug
tolerance
barbiturates
physical dependence
depressant
stimulant
cirrhosis
opiate
THC
carbon monoxide
withdrawal
drug
drug abuse
crack
nicotine

Down
1 a poison found in the leaves of tobacco
2-16 drug that requires a doctor's
    prescription (2 words)
3 loss of liver function caused by alcohol
    abuse
4-12 effect of drug abuse in which the body
    cannot function properly without the drug
5 a person who is physically dependent on
    drugs is suddenly taken off that drug goes
    through
6-15 a poisonous gas that is given off when
    tobacco burns (2 words)
7 belong to the group of drugs called
central nervous system depressants
8 drug that increases the activity of the synpathetic
nervous system and produces a sense of euphoria,
well-being, or increased alertness.
11 highly addictive drug derived from morphine

Drugs, Alcohol, and Tobacco

Across
3 an extremely dangerous form of cocaine
    that is smoked
9 drug that slows down the actions of the
    nervous system
10 substance that has an effect on the body
13 effect of drug abuse in which a person
    must take more of a drug each time to
    get the same effect
17 pain killing drug produced from the
    opium poppy
18 a chemical in marijuana that effects
    people in different way

31

Genetics

Word List

genotype
cross pollination
traits
gene
phenotype
dominant
DNA
hypothesis
incomplete dominance
hybrid
Gregor Mendel
nitrogen base
recessive
replication
genetics

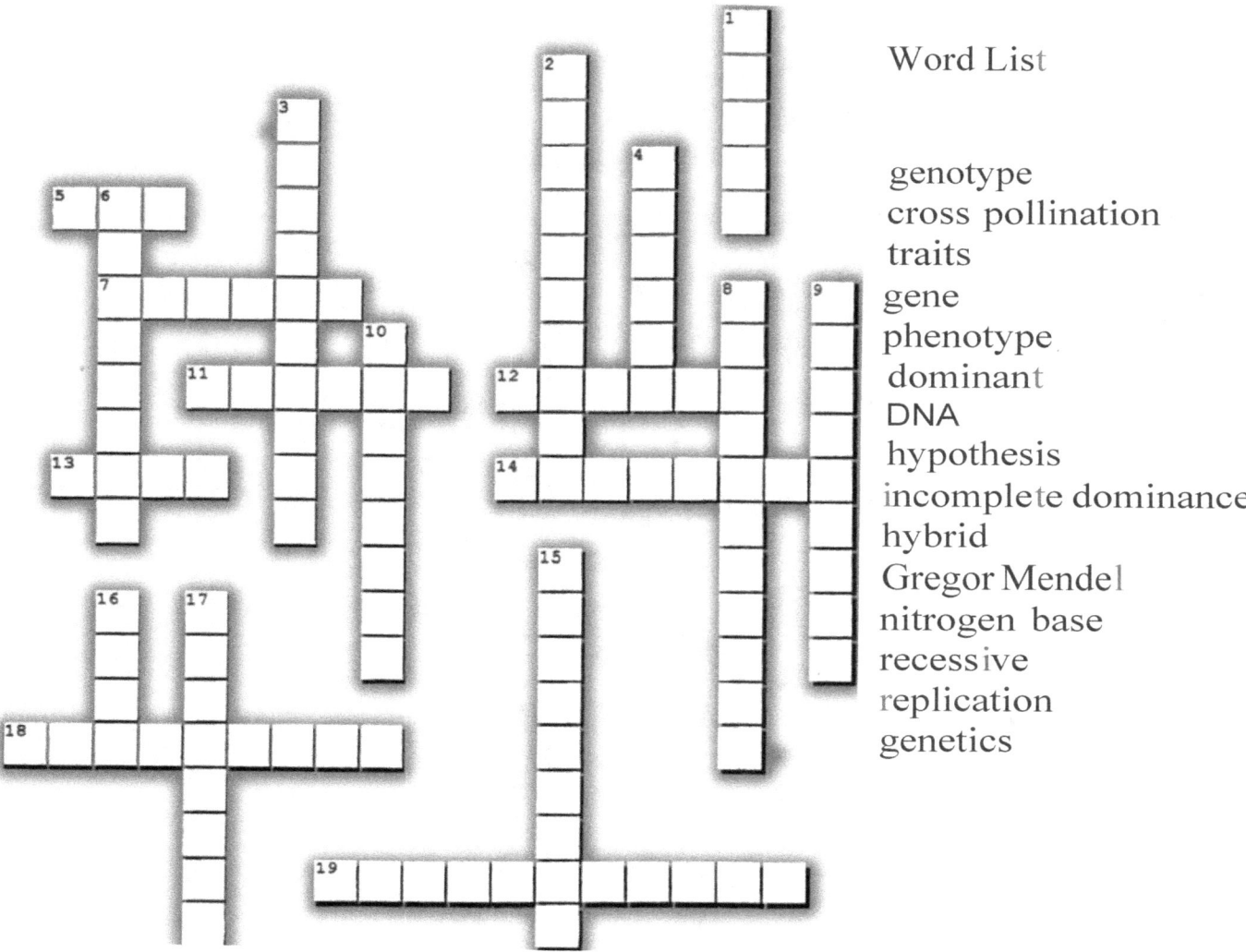

Genetics

Across

5 the basic substance of heredity
7 characteristics of an organism
11-12 known as the Father of Genetics
13 basic unit of heredity
14 study of heredity
18 visible characteristic of an organism
19 the process of making more DNA

Down

1-8 process in which a plant pollinates itself
2-15 condition that occurs when a gene is
      neither dominant nor recessive
3 suggested solution to a problem
4 organism With 2 different genes for a
   particular trait
6-16 substance in DNA that contains the
      element nitrogen
9 weaker trait in genetics
10 stronger trait in genetics
17 genetic makeup of an organism

Genetics & Genetic Engineering

Word List

inbreeding
hybrid
color blindness
Downs Syndrome
sex linked traits
hybridization
meiosis
allele
plasmid
amniocentesis
hybrid vigor
recombinant DNA
genetic engineering
biotechnology
multiple allele

Down

1 crossing of two genetically different but related species of an organism

2 process of removing fluid from the sac surrounding a developing baby

3 bacterial DNA in the form of a ring

4-13 ability of a hybrid to grow faster or larger than their parents 2 words)

5-19 new piece of DNA produced by combining parts of separate DNA strands (2 words)

6-16 more than two alleles that combine to determine a certain characteristic (2 words)

7-17 process in which genes. or parts of DNA are transferred from one organism into another organism (2 words)

9 application of technology to the study and solution of problems involving living things

10-15 a condition that is caused by nondisjunction of the 21st chromosome pair (2 words)

12-14-20 characteristic passed from parent to child on a sex chromosome Q words)

18 breeding that involves crossing plants or animals that have the same or very s1m11.ar sets of genes

Across

8-9 sex-linked trait that causes the inability to distinguish between certain colors (2 words)

11 the process that produces sex cells

21 each member of a gene pair

22 organism with two different genes for a particular trait

23 the process of making more DNA

The Theory of Evolution

natural selection
gradualism Alfred
Wallace
molecular clock
radioactive dating
anatomy
embryology
fossil record
relative
dating
homologous
half-life
adaptation
primate
fossil
evolution

Across

1 s1rnilar in structure
4 order of mammals that include humans.
  apes, and monkeys
8-9 time it lakes for haJf of a radioactive element
  to decay (2 words)
10 study of the structure of living things
15 tmprint or remains of plants or an1mals
  that existed in the past
17-18 most complete biological record of life
  on earth (2 words)
19-20 method based on radioactive elements
  used by scientists to measure the age of
  fossils or the age of the rocks in which fossils
  arc found (2 words)
21-22 survival and reproduction of those
  organisms best adapted to their surrounding

Down

2 belief that evolution is a slow and steady
  process
3-13 scale used to estimate the rate of change
  in proteins over time (2 words)
5-11 worked with Darwin to develop the theory
  of evolution (2 words)
6 study of developing organisms
7-16 method of determining the age of fossils
  that involves comparing the rock layers in
  which the fossils are formed (2 words)
12 change that increases an organism's chances
  of survival
14 change in a species over time

Energy and Living Resources

Word List
nuclear energy
solar energy
extinction
hydroelectric plant
petroleum
alternative
geothermal energy
nuclear fusion

Down

1-13 plant that uses energy from moving water to produce electricity (2 words)

2 energy source includes solar, water and wind, geothermal and nuclear

3-10 process by which atoms are combined and energy is released (2 words)

4-11 species in danger of becoming extinct (2 words)

5-12 energy that comes from heat created in the earth (2 words)

8 process by which a species passes out of existence

9-14 energy from the sun (2 words)

Across

6-7 energy located within the nucleus of an atom

13 liquid fossil fuel

Ecosystem Relationships

Down

1 symbiotic relationship that is helpful to both organisms
2 struggle among living things to get the proper amount of food, water, and energy
3-13 process of gradual change within a community (2 words)
4 relationship that exists between a predator and its prey
5 symbiotic relationship in which only one partner in the relationship benefits
6 organism that feeds on other living organisms
7 symbiotic relationship in which one organism is harmed by the other organism
10 organism in which another organism lives
11 animal that is killed and eaten by a predator.

Across

8-9 living or non-living factor in an environment that can stop a population from increasing in size
12 animal that kills and eats other animals
14 a relationship in which one organism lives on, near, or even inside another organism

36

Biomes

Word List

grassland
taiga
savanna
coniferous forest
climate
tundra
desert
biome
permafrost
tropical rain forest
phytolankton
marine biome
canopy

Across

2-4-5 forest biome that receives at least
   200 centimeters of rain yearly (3 words)
9 biome that receives less than 25 centimeters
   of rainfall yearly
11 grassland
12 division of area with similar climate, plants,
   and animals
14 northernmost area of a coniferous forest
   biome
16 roof formed by tall trees in the forest
19 average weather in a particular place over
   a long period of time

Down

3-13 northernmost forest biome, which
   contains conifers, or cone-bearing trees
   (2 words)
6 microscopic plants that live on the surface
   of the ocean
7 biome that rims the Arctic Ocean around
   the North Pole and has a cold, dry climate
8-17 biome that contains freshwater lakes,
   ponds, swamps, streams, and rivers (2 words)
10-18 ocean biome (2 words)
15 biome made up mainly of grasses that
   receive between 25 and 75 centimeters
   of rainfall yearly

Living Things, Food and Energy

Word List

food chain
herbivore
decompose
photosynthesis
biotic factor
ecosystem
food web
consumer
omnivore
autotroph
environment
hetertroph
producer
carnivore
niche
abiotic factor
habitat
ecology
population

Across

1-2 includes all the food chains in an
  ecosystem that are connected together
  (2 words)
4 the place in which an organism lives
5 organism that eats only plants
6 organism that can make its own food
13 organism that eats both plants and
  animals
16 role of an organism in its community
  or environment
17 group of organisms in an area that
  interact with on another, together with
  their non-living environment
20 study of relationships and interactions
  of living things with one another and
  with their environment
22 group of the same type of organism
  living together in the same area

Down

3 all the living and non-living things
  with which an organism interacts
4 organism unable to make its own food
7 organism that feeds on dead organic matter
  and breaks down into simpler substances
8 a flesh eater
9-18 living things in an environment
  (2 words)
10-19 non-living things in an environment
  (2 words)
11 process by which plants make their own
  food
12 an organism that feeds directly or indirectly
  on producers
14 organism that can make its own food from
  from simple substances
15-21 food and energy links between the
  different plants and animals in an ecosystem

38

Soil, Air and Water Conservation

Word List

depletion
hazardous waste
windbreak
pollution
pesticide
temperature inversion
erosion
topsoil
renewable
conservation
groundwater
nonrenewable
topsoil
acid rain
fossil fuel
litter
contour farming
strip cropping
smog
terracing

Across

6 thick cloud of pollutants
11 process in which soil is carried off by water or wind
12-13 rain containing nitric acid and sulfuric acid (2 words)
14 rich upper layer of soil
22 poisonous
24 farming method in which a slope is made into a series of level plots in a strip-like pattern to avoid erosion
25 process in which nutrients are washed away from the soil by water
26 natural resource what cannot be replaced by nature

Down

1-17 waste that burns easily, is poisonous, or reacts dangerously with other substances (2 words)
2 introduction of harmful substances into the environment
3-20 atmospheric condition in which a layer of cool air containing pollutants is trapped near the ground under a layer of warm air (2 words)
4-15 product of decayed plants and animals that is preserved in the earth's crust over millions of years (2 words)
5 natural resource that is replaced by nature
7 underground water
8-18 farming method in which a slope is plowed horizontally across its face to avoid erosion (2 words)
9 wise use of natural resources so that they will not be used up too soon or used in a way that will damage the environment
10-16 farming method in which strips of cover crops are grown between strips of other crops to hold down the soil (2 words)
19 row of trees planted between fields of crops to prevent erosion due to wind
21 chemical used to kill harmful insects or other pests
23 material that is disposed of in improper place

SCIENCE

```
D  I  D  H  M  P  D  G  F  C  U  M  M  V  Y
A  S  A  L  I  Y  H  U  G  T  I  H  I  T  R
T  Y  W  Z  L  G  B  I  X  C  O  O  C  U  Z
A  H  K  I  K  E  S  N  R  H  E  B  R  Z  T
P  S  I  S  E  H  T  O  P  Y  H  P  O  D  F
S  E  X  T  K  G  B  K  G  T  M  P  O  A  H
X  J  L  S  H  I  M  O  Y  R  T  R  R  P  H
K  M  Z  B  O  E  L  S  S  L  C  W  G  K  V
I  N  J  L  A  O  O  K  R  O  V  U  A  F  V
N  H  O  J  C  I  D  R  X  R  U  E  N  S  M
N  G  W  E  G  F  R  K  Y  T  E  X  I  V  E
Y  H  I  G  U  C  G  A  P  N  S  R  S  W  E
W  B  O  T  A  N  Y  U  V  O  Z  V  M  O  D
W  L  T  J  Z  Z  Q  N  K  C  I  H  W  R  S
U  T  D  H  L  V  D  Z  I  L  V  P  Y  M  W
```

BOTANY   CONTROL   DATA
ECOLOGY   HYPOTHESIS   MICROBIOLOGY
MICROORGANISM   THEORY   VARIABLE

SCIENTIFIC MEASUREMENTS

```
C  U  C  T  X  K  K  N  Q  K  U  Q  R  N  O
E  S  A  N  F  A  K  I  C  Z  D  B  O  E  D
L  E  E  E  M  M  I  L  L  I  L  I  T  E  R
S  J  K  M  V  A  Q  Y  W  O  S  X  N  D  L
I  V  A  E  U  E  S  N  I  R  G  S  T  N  G
U  C  U  R  N  L  W  S  E  N  I  R  V  W  Q
S  Z  J  U  O  W  O  V  P  T  C  C  A  K  U
J  X  G  S  O  B  N  V  Y  B  O  U  K  M  F
N  R  M  A  X  O  Y  T  I  V  A  R  G  A  U
O  V  D  E  C  I  Y  H  W  T  B  Q  U  G  E
Q  W  B  M  T  Q  S  G  M  E  T  E  R  P  N
U  U  W  N  F  F  H  I  J  Z  T  S  L  A  Y
I  Y  E  M  U  M  W  E  B  K  S  G  B  I  L
P  C  K  Y  U  L  K  W  W  R  X  A  F  B  V
I  K  P  C  F  U  K  B  H  C  H  H  F  M  V
```

CELSIUS	CENTI	CONVERSION
DENSITY	GRAVITY	KILOGRAM
MASS	MEASUREMENT	METER
MILLILITER	VOLUME	WEIGHT

Characteristics of Living Things

```
W   M   A   S   E   X   U   A   L   L   N   L   Q   K   N
W   E   F   O   Y   N   W   B   J   S   O   S   X   Z   O
D   T   Z   W   L   H   L   Q   U   Z   I   N   O   E   I
F   A   D   O   A   T   E   L   S   T   T   O   D   D   T
X   B   E   E   U   R   X   A   K   I   O   I   H   S
T   O   C   X   D   M   M   T   Z   S   T   G   M   F   E
Z   L   E   Y   I   O   S   B   N   G   E   C   C   I   G
J   I   Q   T   E   O   O   E   L   S   P   O   U   U   N
S   S   S   E   E   K   T   L   T   O   M   R   D   O   I
L   M   F   M   T   A   E   I   B   U   O   A   Y   A   X
V   K   O   E   E   Z   O   G   X   D   C   D   T   J   M
W   H   W   C   U   N   A   T   X   K   L   B   E   O   E
R   E   S   P   I   R   A   T   I   O   N   O   J   D   M
N   O   I   T   C   U   D   O   R   P   E   R   C   E   O
N   O   I   T   E   R   C   X   E   D   A   P   G   L   C
```

ASEXUAL	ATOM	COLDBLOODE
COMPETITION	DIGESTION	EXCRETION
HOMEOSTATIS	INGESTION	METABOLISM
REPRODUCTION	RESPIRATION	STIMULUS
WARMBLOODED		

Chemistry of Living Things

```
H   W   M   L   R   F   D   N   I   H   K   B   K   T   R
D   H   W   E   U   M   O   O   S   B   Y   E   X   F   A
X   N   T   J   X   V   C   R   F   I   L   B   E   J   G
N   A   U   T   U   N   N   T   M   U   Z   R   N   N   U
W   O   N   O   F   Z   B   C   C   U   R   E   E   Q   S
X   K   T   H   P   W   J   E   J   U   L   J   U   F   D
C   R   D   O   A   M   L   L   M   W   C   A   T   O   M
T   X   J   P   R   O   O   E   Q   J   T   J   R   T   Z
K   D   N   Q   M   P   S   C   P   L   G   B   O   F   R
N   K   E   E   E   L   E   M   E   N   T   N   N   V
O   J   C   Y   X   G   Q   S   I   Y   Z   R   V   P   R
X   R   W   R   E   T   T   A   M   W   Q   S   K   N   Z
D   A   N   Z   K   L   K   V   O   D   E   K   N   O   H
A   W   I   D   M   S   D   L   H   D   K   G   G   J   S
X   D   O   R   J   U   X   I   W   G   P   E   U   M   M
```

ATOM	COMPOUND	ELECTRON
ELEMENT	FERRUM	FORMULA
MATTER	MOLECULE	NEUTRON
PROTON	SUGAR	WATER

Cell Structure and Function

```
C U M W X K T Y V K T R L R O
A H M S E K T V E S R P L L R
S I R B A C F L A P R S Y I G
U S R O Q L L L R I C U H M A
E Y Z D M E P O B Q D L P E N
L G H O N O T O F S C O O T I
C U V A R O S E T A G E R A S
U T G O P O H O P Y F L O B M
N R L L M M A C M I C C L O V
O H A E U V B I O E T U H L A
C S R A E L C U N C B N C I C
M D H O M E O S T A T I S S U
E S O L U L L E C A D I A M O
L Y S O S O M E F Y D V M S L
P R O T E I N D W W U H F P E
```

CELLULOSE	CHLOROPHYLL	CHLOROPLAST
CHROMOSOME	CYTOPLASM	HOMEOSTATIS
LYSOSOME	METABOLISM	MITCOCHONDRIA
NUCLEAR	NUCLEOLUS	NUCLEUS
ORGANELLE	ORGANISM	PROTEIN
PROTOPLASM	RIBOSOME	VACUOLE

Cell Processes

```
N  H  O  T  U  F  L  Q  C  C  L  N  N  E  J
E  O  E  S  A  H  P  A  T  E  M  O  U  S  B
M  G  I  P  M  U  Q  E  C  N  K  I  P  A  S
O  E  G  T  Q  O  S  F  O  T  O  T  V  H  A
E  T  T  F  C  A  S  I  R  R  P  A  W  P  N
A  S  A  A  H  U  S  I  C  I  R  R  A  A  Y
E  R  A  P  B  S  D  H  S  O  A  I  A  N  S
V  O  O  H  U  O  R  O  X  L  C  P  V  A  E
F  R  R  F  P  O  L  U  R  E  N  S  X  A  Z
P  N  F  N  M  O  R  I  Y  P  J  E  B  W  U
M  I  Y  A  Y  W  L  S  S  L  E  R  L  Z  F
D  X  T  Z  Y  F  U  E  E  M  Z  R  F  M  X
P  I  M  W  S  T  M  I  T  O  S  I  S  L  S
N  I  N  T  E  R  P  H  A  S  E  B  W  P  X
S  Q  M  F  E  H  S  A  N  H  Q  R  Q  Z  W
```

ANAPHASE
DIFFUSSION
METAPHASE
PROPHASE
TELOPHASE

CENTRIOLE
INTERPHASE
MITOSIS
REPRODUCTION

CHROMATIN
METABOLISM
OSMOSIS
RESPIRATION

Classification of Living Things

```
F  X  E  V  N  B  N  H  A  A  S  T  B  J  R  A  I  T  B  R
A  O  U  G  S  M  O  R  Q  U  P  U  W  R  R  Q  A  D  A  M
M  B  X  T  N  C  I  P  N  X  T  X  E  I  H  X  H  L  D  O
I  S  Q  I  H  U  T  C  B  V  I  O  S  A  O  I  U  T  S  D
L  T  V  K  H  M  A  G  H  L  G  T  T  N  N  L  J  C  P  G
Y  P  G  C  U  X  C  E  N  P  O  Y  O  R  L  N  Z  S  E  N
P  I  X  L  A  K  I  T  B  M  M  U  E  O  R  I  O  C  I
N  E  Y  F  F  X  F  G  L  V  Y  C  C  O  T  P  X  L  I  K
G  H  O  G  M  H  I  E  S  C  W  I  K  T  Q  K  H  L  E  A
P  F  R  Z  Z  P  S  T  H  F  N  M  O  N  E  R  A  N  S  B
L  S  D  T  N  O  S  B  Y  U  C  R  K  V  S  W  W  O  K  Q
L  O  E  T  Y  R  A  T  B  B  R  S  U  K  M  C  Q  Y  P  O
B  O  R  I  T  T  L  R  A  L  U  L  L  E  C  I  T  L  U  M
H  P  N  Y  A  O  C  X  D  L  J  J  P  Y  W  H  R  H  B  H
K  H  X  K  L  R  G  V  I  M  X  X  D  E  T  C  Z  V  H  X
G  F  A  L  U  E  K  D  I  R  F  U  J  U  P  R  V  D  L  P
S  Z  S  L  O  T  Q  N  K  C  Z  Q  L  T  X  O  A  N  J  C
N  I  T  A  L  E  Y  G  A  A  R  U  F  H  U  X  T  A  O  A
P  J  A  G  E  H  X  D  D  G  Z  G  W  B  R  U  M  S  S  G
V  O  Q  J  C  E  N  N  J  L  Q  N  N  V  L  O  I  Y  W  I
```

ARISTOTLE	AUTOTROPH	CLASSIFICATION
FAMILY	HETEROTROPH	KINGDOM
LATIN	LINNAEUS	MONERAN
MULTICELLULAR	ORDER	PHYLUM
SPECIES	TAXONOMY	UNICELLULAR

46

Viruses and Bacteria

```
Q F E N C B I F G R Y S P N O
J C L G S N L V E H R N O H N
O I Y D A A P T D J B I Y O W
A V W T G H C L A A T X P S U
J K I E O A P N N A R O A T B
Y E L R B P J O Z W I T R F D
G L L K U H L I I V N T A Q Y
A B Z T H S R A Z E J K S J L
T U R E R U X A S H F Z I B T
E E N O E N D T Q M E R T R H
B A N T I B I O T I C S E S X
U F S L E C P V U A N H S X N
Y A R G H M L S Q G D C L G B
P T A I R E T C A B P S F X M
E P Z Z Z O F H C O C C I F M
```

ANTIBIOTICS

COCCI

HOST

PASTEURIZATION

BACTER

CYTOPLASM

IOPHAGE

TOXINS

BACTERIA

FLAGELLA

PARASITES

VIRUS

Protozoans

```
S  U  E  L  C  U  N  O  R  C  I  M  N  S  C
T  N  N  M  E  D  F  E  V  M  L  A  S  U  O
K  S  Q  A  F  U  T  H  U  F  R  V  Z  E  N
D  J  A  P  O  I  G  I  I  E  K  N  P  L  J
Y  C  Y  L  S  Z  C  L  F  F  A  X  L  C  U
T  D  I  A  P  E  O  I  E  O  R  E  A  U  G
M  O  R  L  M  O  C  T  Z  N  T  T  S  N  A
S  A  P  A  I  I  R  O  O  E  A  A  M  O  T
P  P  R  S  M  A  R  O  L  R  B  L  O  R  I
D  A  O  A  E  O  T  L  L  K  P  L  D  C  O
P  G  R  R  P  Y  U  E  G  A  L  E  I  A  N
A  O  V  S  E  G  E  G  G  K  B  G  U  M  W
F  D  O  P  O  D  U  E  S  P  C  A  M  B  C
E  N  I  D  O  C  R  A  S  O  A  L  V  S  G
E  L  C  I  L  L  E  P  Q  A  C  F  I  S  C
```

BALOROPLAST	CILIATE	CONJUGATION
EUGLENA	EYESPOT	FLAGELLATE
FORAMICIFERAN	GULLET	MACRONUCLEUS
MICRONUCLEUS	PARAMECIUM	PARASITE
PELLICLE	PLASMODIUM	PROTOZOAN
PSEUDOPOD	SARCODINE	SPORE
SPOROZOAN		

Algae and Fungi

```
R  P  T  V  I  I  A  Y  N  K  S  A  B  W  P  F  Y  V  K  O
U  U  P  A  Q  W  X  N  W  D  O  U  I  H  Q  U  E  R  R  N
W  J  T  T  G  B  Q  F  E  Z  L  N  O  T  K  N  A  L  P  F
U  U  R  A  L  U  L  L  E  C  I  T  L  U  M  G  S  E  I  Z
J  Q  B  X  H  D  X  R  M  R  O  N  U  D  A  I  T  P  U  O
S  G  F  K  V  D  A  R  A  S  M  G  M  U  G  A  S  N  T  J
N  G  V  P  V  I  M  O  Y  L  M  E  I  F  L  P  V  I  W  A
K  E  W  Q  X  N  O  N  X  B  U  G  N  L  A  W  W  Z  N  L
H  I  L  P  F  G  T  W  K  J  G  C  E  T  L  B  E  L  C  L
J  B  H  Y  P  H  A  E  Y  C  R  G  S  B  A  A  E  J  S  A
M  Q  H  S  E  T  I  S  A  R  A  P  C  A  G  T  E  U  Q  F
N  K  E  S  U  Q  D  D  R  L  L  R  E  P  V  N  I  G  M  N
T  O  I  G  L  C  S  L  F  K  U  B  N  N  S  N  H  O  H  K
I  S  L  L  Y  H  P  O  R  O  L  H  C  P  B  L  O  L  N  Z
Y  T  A  B  Q  R  N  M  S  A  L  A  E  T  F  R  F  N  P  E
B  Q  R  F  C  I  D  C  D  K  E  P  T  N  H  L  C  O  A  K
L  F  Z  T  D  U  I  D  Q  C  C  D  G  S  F  P  Y  I  X  A
P  U  K  D  S  L  E  N  I  F  I  D  U  J  M  Y  S  O  M  M
U  K  S  S  E  R  O  P  S  N  N  M  G  L  X  V  S  F  O  E
A  M  K  E  S  X  L  H  S  Q  U  K  J  A  Y  D  S  V  J  R
```

AGAR
ALGA
BIOLUMINESCENCE
BLADDERS
BUDDING
CHLOROPHYLL
CLUB
DIATOM
DINOFLAGELLATE

FERMENTATION
FUNGI
GILL
HOLDFAST
HYPHAE
MOLDS
MULTICELLULAR
MUSHROOM

NONVASCULAR
PARASITES
PHOTOSYNTHESIS
PLANKTON
SPORE
STALK
UNICELLULAR
YEASTS

Vascular Plants

```
G D N O R F G V Z P E N N L G
E S E M O Z I H R S D O N Y E
R R T M A R S C O S I K M P L
M C E J U I Q C V T M N Q H C
I T V P M I U X A F O I R L I
N K J P R L B Z T S P R E O T
A P L N G O I M P Y F O E E U
T E O I D L D E A L R S R M C
I F R I I Y R U C C N A Q P O
O B I T Q M H S C H F I V N L
N X R L S T O M A T A X B O Z
K E Y K R D K L S U I M L H R
F F G G D S R D R O U O A P L
G I B B E R E L L I N S N T W
G D U K T N N I D Z Y B K B C
```

CAMBIUM   CUTICLE   FERTILIZATION
FROND   GERMINATION   GIBBERELLINS
GLUCOSE   GYMNOSPERMS   OVARY
OXYGEN   PHLOEM   REPRODUCTION
RHIZOMES   SIMPLE   SORI

Ferns

```
A  A  N  E  R  M  Z  S  U  H  I  O  D  B  X
P  Y  W  T  S  V  V  N  U  R  W  Y  D  W  R
J  L  M  Y  N  O  E  K  A  D  Z  O  A  W  M
M  K  S  H  D  M  R  S  N  T  R  F  H  W  W
X  N  L  P  N  O  P  I  O  X  N  U  T  C  C
T  A  X  O  O  O  W  R  I  N  O  B  J  V  P
I  U  S  T  R  R  H  H  L  A  Z  R  G  J  Y
N  X  S  E  F  A  O  T  Y  B  G  H  W  M  D
Q  Y  S  M  X  L  Z  P  D  D  A  I  R  H  X
C  X  H  A  Y  U  W  S  H  F  M  Z  X  V  M
W  O  D  G  A  C  A  I  N  Y  E  O  I  T  M
B  Y  P  D  Y  S  G  L  G  R  T  M  X  O  S
D  J  X  O  X  A  N  O  S  V  E  E  F  M  D
U  N  Y  T  J  V  T  U  C  B  S  F  C  S  K
W  H  I  I  H  W  O  F  E  Y  M  F  E  J  F
```

ASEXUAL	FERNS	FROND
GAMETES	GAMETOPHYTE	RHIZOME
ROOTS	SORI	SPORE
SPOROPHYTE	VASCULAR	WIND

Seed Plants

```
X  X  B  X  Y  C  H  A  B  R  H  D  U  S  B
R  M  K  Y  G  U  F  U  I  X  R  D  I  E  T
C  S  T  L  N  T  L  N  Q  A  M  S  A  C  C
S  H  P  E  O  I  G  W  U  R  E  N  N  E  H
I  R  L  M  P  C  J  G  E  H  G  H  O  S  L
M  W  X  O  S  L  M  P  T  I  U  S  T  U  O
R  R  Q  B  R  E  S  N  O  I  U  O  R  F  R
E  D  E  E  S  O  Y  S  C  H  M  U  F  H  O
D  T  R  A  N  S  P  I  R  A  T  I  O  N  P
I  F  S  Y  O  E  M  H  T  M  M  T  O  V  L
P  W  G  T  R  G  D  A  Y  W  E  B  E  W  A
E  R  O  M  T  U  B  E  R  L  O  O  I  B  S
B  H  P  A  L  I  S  A  D  E  L  O  L  U  T
P  S  U  O  E  C  A  B  R  E  H  O  D  H  M
N  E  G  Y  X  O  P  O  H  D  Q  K  B  Y  P
```

ANGIOSPERM	CAMBIUM	CHLOROPHYLL
CHLOROPLAST	CUTICLE	EPIDERMIS
GUARD	GYNOSPERM	HERBACEOUS
OXYGEN	PALISADE	PHLOEM
PHOTOSYNTHESIS	RING	SEED
SPONGY	STOMATA	TRANSPIRATIO
TUBER	WOODY	XYLEM

Gymnosperms and Angiosperms

```
P  R  Y  R  A  V  O  B  W  C  N  A  L  S  N
S  O  E  P  E  Q  L  F  O  O  O  N  I  T  O
R  T  L  W  E  X  R  T  I  A  T  G  T  I  I
W  K  Y  L  O  U  Y  T  C  B  L  I  S  G  T
G  C  V  L  I  L  A  G  Z  L  Q  O  I  M  A
E  M  Q  T  E  N  F  B  A  G  R  S  P  A  Z
B  Z  B  D  I  N  A  T  S  S  E  P  A  L  I
J  Y  O  M  Y  Q  E  T  N  B  D  E  O  M  L
I  N  R  I  K  P  S  I  I  L  B  R  A  U  I
C  E  F  S  R  E  I  F  N  O  C  M  Q  N  T
G  S  T  A  M  E  N  D  W  P  N  E  E  O  R
G  Y  M  N  O  S  P  E  R  M  S  L  H  W  E
N  J  S  M  J  X  U  I  W  P  L  U  K  P  F
Z  T  T  N  O  D  J  E  Y  O  C  V  Q  N  Y
J  C  M  D  T  M  U  B  P  R  T  O  K  H  Y
```

ANGIOSPERM	CONFIERS	COTYLEDON
FERTILIZATION	FLOWER	FRUIT
GERMINATION	GYMNOSPERMS	OVARY
OVULE	PETAL	PISTIL
POLLEN	POLLINATION	SEPAL
STAMEN	STIGMA	STYLE

Invertebrates

```
N E M A T O C Y S T W N X Y E
K M E X O S K E L E T O N H L
H H T R I C H I N O S I S O T
P L A T Y H E L M I N T H S N
E A M D O P O R H T R A U T A
G P O R I F E R A N Y Z Z B M
A U R E G E N E R A T I O N A
L P P G S E D E P I L L I M R
F P H E R O M O N E S I S P A
O P O R C S E D E P I T N E C
M W S N A E C A T S U R C Y H
A F I E T A R B E T R E V N I
C H S I F Y L L E J K F N C N
P O R E S D I L E N N A Z X D
C O R A L S N E M A T O D E S
```

ANNELIDS

CAMOFLAGE

CROP

FERTILIZATION

JELLYFISH

MILLIPEDES

PHEROMONES

PORIFERAN

TRICHINOSIS

ARACHINDS

CENTIPEDES

CRUSTACEANS

HOST

MANTLE

NEMATOCYST

PLATYHELMINTHS

PUPA

ARTHROPOD

CORALS

EXOSKELETON

INVERTEBRATE

METAMORPHOSIS

NEMATODES

PORE

REGENERATION

More about Invertebrates

```
E  C  N  O  T  E  L  E  K  S  O  X  E  F  J
T  R  M  I  L  L  I  P  E  D  E  S  L  M  E
A  U  D  R  A  Z  Z  I  G  B  B  A  S  K  L
R  S  E  R  S  E  D  E  P  I  T  N  E  C  L
B  T  B  E  P  T  L  B  V  Y  Q  L  G  L  Y
E  A  K  O  A  F  C  A  H  Q  P  L  N  H  F
T  C  R  E  G  E  N  E  R  A  T  I  O  N  I
R  E  R  A  D  U  L  A  S  V  L  Y  P  Z  S
E  A  B  F  U  M  O  J  Q  N  A  D  S  S  H
V  N  N  O  I  T  A  Z  I  L  I  T  R  E  F
N  S  E  N  O  M  O  R  E  H  P  R  P  T  L
I  L  T  A  R  T  H  R  O  P  O  D  Q  A  L
X  H  T  R  I  C  H  I  N  O  S  I  S  E  K
S  E  D  O  T  A  M  E  N  U  A  O  O  I  V
S  T  A  R  F  I  S  H  Z  X  O  K  S  U  F
```

ARTHROPOD	CENTIPEDES	CRUSTACEANS
EXOSKELETON	FERTILIZATION	FLATYHELMINTHS
GIZZARD	INSECTS	INVERTEBRATE
JELLYFISH	LARVA	MILLIPEDES
NEMATODES	PHEROMONES	PORE
RADULA	REGENERATION	SETAE
SPONGES	STARFISH	TRICHINOSIS

Coldblooded Vertebrates

```
T  L  O  M  V  E  R  T  E  B  R  A  E  P  T
A  M  P  H  I  B  I  A  N  X  H  G  U  E  H
E  X  T  E  R  N  A  L  V  Q  E  W  R  Q  N
A  J  Y  Z  W  R  E  P  T  I  L  E  S  O  D
N  O  T  E  L  E  K  S  O  D  N  E  I  C  E
B  B  L  Z  R  F  G  M  C  E  N  T  S  A  D
F  A  V  A  O  E  O  N  T  Q  A  Y  E  R  O
R  C  E  R  N  N  D  A  A  N  J  Z  L  T  O
S  K  S  I  E  R  D  N  R  F  A  E  O  I  L
G  B  N  V  P  R  E  E  A  I  A  P  P  L  B
L  O  Y  G  O  Z  B  T  B  M  S  A  D  A  M
P  N  Q  H  Y  I  W  Y  N  M  A  B  A  G  R
Q  E  C  J  H  H  S  I  F  I  H  L  T  E  A
D  E  D  O  O  L  B  D  L  O  C  T  A  S  W
R  E  D  D  A  L  B  M  I  W  S  U  E  S  A
```

AMPHIBIAN	BACKBONE	CARTILAGE
CHORDATE	COLDBLOODED	ENDOSKELETON
EXTERNAL	FANG	FISH
HIBERNATION	INTERNAL	MOLT
REPTILES	SALAMANDER	SWIMBLADDER
TADPOLES	VENOM	VERTEBRAE

Warmblooded Vertebrates

```
C  T  F  E  M  B  R  X  L  I  W  H  E  M  E
A  S  E  B  R  U  I  A  J  A  K  E  R  A  N
S  T  A  R  O  O  I  R  R  B  M  D  O  M  I
C  T  N  T  R  P  V  M  D  I  F  G  V  M  N
S  A  N  E  U  I  B  I  G  S  G  E  I  A  A
G  O  S  S  C  L  T  R  N  O  N  H  B  R  C
C  B  R  R  O  A  A  O  S  R  P  O  R  Y  W
M  A  J  O  I  T  L  M  R  O  A  G  E  S  H
M  E  D  Z  E  A  W  P  B  Y  Y  C  H  R  A
Y  E  O  L  L  I  D  A  M  R  A  O  N  E  L
D  E  T  A  B  U  C  N  I  A  N  O  R  V  E
I  N  C  I  S  O  R  A  B  B  I  T  N  A  P
Z  P  E  T  A  M  I  R  P  J  G  L  Z  E  Y
O  Q  M  P  Q  X  M  K  A  L  A  P  G  B  M
N  W  O  D  Z  T  M  O  N  O  T  R  E  M  E
```

AIRSAC	ARMADILLO	BATS
BEAVERS	BIRDS	CANINE
CARNIVORE	CONTOUR	DOWN
HEDGEHOG	HERBIVORE	INCISOR
INCUBATE	MAMMARY	MARSUPIAL
MIGRATE	MONOTREME	PLACENTA
PRIMATE	RABBIT	TERRITORY
WARMBLOODED	WHALE	

The Skeletal and Muscular System

```
O  N  C  C  A  S  R  U  M  E  F  L  N  L  C
C  R  P  A  L  L  M  S  E  S  U  I  A  A  O
A  W  G  D  L  A  L  O  F  Y  F  G  I  T  N
R  O  T  A  Q  C  V  E  O  R  W  A  S  E  N
D  R  X  N  N  E  I  I  T  T  S  M  R  L  E
I  R  M  T  I  S  G  U  C  A  H  E  E  E  C
A  A  Q  K  Q  O  Y  A  M  L  P  N  V  K  T
C  M  N  Q  X  B  J  S  L  I  E  T  A  S  I
O  S  S  I  F  I  C  A  T  I  O  N  H  R  V
T  A  D  G  U  E  U  H  N  E  T  V  S  J  E
E  K  V  E  R  T  E  B  R  A  M  R  H  D  V
N  X  P  H  A  L  A  N  G  E  S  J  A  Z  R
D  R  M  L  I  S  U  R  E  M  U  H  Q  C  E
O  D  D  A  M  U  E  T  S  O  I  R  E  P  N
N  E  L  B  I  D  N  A  M  U  S  C  L  E  S
```

CALCIUM	CARDIAC	CARTILAGE
CLAVICLE	CONNECTIVE	EPITHELIAL
FEMUR	HAVERSIAN	HUMERUS
JOINT	LIGAMENT	MANDIBLE
MARROW	MUSCLE	NERVE
ORGANSYSTEM	OSSIFICATION	PATELLA
PERIOSTEUM	PHALANGES	SKELETAL
SMOOTH	TENDON	VERTEBRA

The Digestive System

```
J  D  X  S  B  P  R  M  C  V  G  W  L  C  L
P  H  I  I  A  O  E  H  O  A  U  A  S  A  A
H  T  L  G  S  E  E  P  L  L  C  E  N  R  R
N  E  Y  I  E  M  R  L  S  I  A  I  I  B  G
A  O  C  A  I  S  B  C  N  I  M  R  C  O  E
V  N  I  C  L  L  T  A  N  A  N  O  S  H  D
I  I  A  T  A  I  H  I  T  A  J  L  I  Y  I
L  L  L  D  P  C  N  I  O  S  P  A  T  D  G
A  H  D  L  E  R  V  V  X  N  I  C  T  R  E
S  E  A  M  I  N  O  A  C  I  D  S  O  A  S
R  S  I  S  L  A  T  S  I  R  E  P  L  T  T
T  N  E  I  R  T  U  N  B  C  Y  J  G  E  I
H  C  A  M  O  T  S  T  E  A  J  A  I  V  V
N  C  A  N  I  N  E  F  V  V  I  T  P  K  E
N  K  L  M  I  N  E  R  A  L  I  V  E  R  T
```

ABSORPTION	AMINOACID	BILE
CALORIE	CANINE	CARBOHYDRATE
CHEMICAL	DIGESTION	DIGESTIVE
EPIGLOTTIS	FECES	GALLBLADDER
INCISOR	LARGE	LIVER
MECHANICAL	MINERAL	MOLAR
NUTRIENT	PANCREAS	PEPSIN
PERISTALSIS	PTYALIN	SALIVA
STOMACH	VILLI	VITAMIN

The Circulatory System

```
A  T  R  I  U  M  E  C  I  P  Y  W  Y  B  A
T  R  A  N  S  F  U  S  I  O  N  R  Q  T  Q
V  N  W  H  Y  N  S  Q  A  K  O  Z  H  M  P
L  E  O  X  E  L  C  I  R  T  N  E  V  A  C
O  T  I  I  T  M  S  F  A  Z  R  N  C  S  A
R  P  E  N  S  E  O  L  I  O  Q  E  P  B  P
E  L  F  L  P  N  U  G  S  B  M  C  Y  C  I
T  A  H  T  E  C  E  C  L  A  R  R  U  I  L
S  S  U  P  R  T  L  T  K  O  E  I  E  N  L
E  M  A  I  M  E  A  E  R  T  B  V  N  O  A
L  A  C  F  R  Y  R  L  R  E  L  I  U  R  R
O  S  P  O  O  U  L  A  P  A  P  C  N  H  Y
H  N  S  A  T  R  O  A  V  R  V  Y  U  C  E
C  I  D  T  M  A  W  W  H  I  T  E  H  X  A
S  R  E  T  E  M  O  N  O  M  G  Y  H  P  S
```

AORTA	ARTERY	ATHEROSCLEROSIS
ATRIUM	CAPILLARY	CHOLESTEROL
CHRONIC	CIRCULATORY	FIBRIN
HEMOGLOBI	HYPERTENSION	LYMPH
PACEMAKE	PLASMA	PLATELET
SEPTUM	SPHYGMONOMETER	TRANSFUSION
VALVE	VEIN	VENTRICLE
WHITE		

The Respiratory and Excretory System

```
J  N  M  N  S  W  L  A  Q  L  E  A  A  L  E
V  A  O  O  I  O  E  L  U  X  Y  L  U  X  E
O  E  Y  I  T  T  J  X  H  G  V  N  C  M  L
C  R  E  A  T  P  R  A  C  E  G  R  G  N  U
A  U  N  L  O  A  L  A  O  E  E  A  O  U  S
L  E  D  I  L  E  R  L  C  T  R  R  B  J  P
C  U  I  V  G  A  U  I  O  H  H  T  D  K  A
O  R  K  E  I  S  Y  R  P  P  E  T  I  S  C
R  E  R  R  P  N  Y  A  E  S  J  A  G  O  B
D  T  K  N  E  S  I  N  K  T  E  B  K  W  N
S  H  F  V  Y  D  E  P  I  D  E  R  M  I  S
D  R  J  S  H  F  Z  N  O  S  T  R  I  L  Y
N  A  T  S  U  H  C  N  O  R  B  N  I  K  S
R  E  D  D  A  L  B  Y  R  A  N  I  R  U  Z
M  L  A  R  Y  N  X  E  L  A  H  N  I  T  S
```

ALVEOLUS	BRONCHUS	CAPSULE
DIAPHRAGM	EPIDERMIS	EPIGLOTTIS
EXCERTION	EXCRETORYSYSTEM	EXHALE
INHALE	KIDNEY	LARYNX
LIVER	LUNG	NEPHRON
NOSTRIL	RESPIRATION	SKIN
TRACHEA	UREA	URETER
URETHRA	URINARYBLADDER	VOCALCORD

The Nervous System

```
I  L  N  D  N  M  M  M  R  V  H  T  S  D  A
N  J  A  M  E  U  E  E  M  X  V  E  X  R  E
K  E  U  R  R  N  C  T  C  U  N  F  E  O  L
M  B  R  D  E  E  D  I  S  S  L  F  L  C  H
Y  O  R  V  P  H  M  R  O  Y  Z  E  F  L  C
U  A  T  O  O  P  R  I  P  S  C  E  A  O
E  V  O  O  N  U  Y  I  L  T  R  T  R  N  C
A  R  C  O  R  N  S  T  R  A  E  O  Q  I  P
B  O  T  A  E  N  R  O  C  E  R  R  S  P  U
E  U  S  U  L  U  M  I  T  S  P  T  Y  S  P
A  N  R  C  E  R  E  B  R  U  M  N  N  E  I
H  O  M  U  L  L  E  B  E  R  E  C  A  E  L
N  G  B  P  M  E  D  U  L  L  A  X  P  K  C
S  N  E  L  R  E  T  I  N  A  O  B  S  S  R
Y  D  H  R  J  H  E  N  Q  N  C  Z  E  R  Q
```

AUTONOMIC	AXON	CENTRAL
CEREBELLUM	CEREBRUM	COCHLEA
CORNEA	DENDRITE	EARDRUM
EFFECTOR	LENS	MEDULLA
MOTOR	NERVOUS	PERIPHERAL
PUPIL	RECEPTOR	REFLEX
RETINA	SENSORYNEURON	SPINALCORD
STIMULUS	SYNAPSE	SYSTEM

The Endocrine System

```
E  J  W  U  P  A  R  A  T  H  Y  R  O  I  D
M  N  T  E  S  T  E  S  X  G  S  R  X  X  S
A  E  D  G  P  E  N  M  P  W  A  Z  B  U
E  N  D  O  C  R  I  N  E  S  Y  S  T  E  M
R  O  J  N  C  V  O  T  I  F  Q  Q  W  T  L
F  M  R  A  Y  R  Y  V  U  X  N  M  E  O  A
Q  R  A  D  H  Q  I  W  A  I  O  S  W  K  H
B  O  D  O  I  F  F  N  L  R  T  R  L  G  T
U  H  R  T  U  Q  U  E  R  Y  A  H  R  O
T  J  E  R  E  E  S  N  O  G  O  M  R  T  P
V  C  N  O  M  N  C  G  H  B  L  R  G  Y  Y
Q  T  A  P  I  P  E  X  Y  Q  Z  A  S  V  H
E  Y  L  I  M  N  U  S  A  E  R  C  N  A  P
P  Z  S  C  S  G  F  N  I  C  V  I  W  D  N
E  N  O  R  E  T  S  O  T  S  E  T  N  R  V
```

ADRENALS	ENDOCRINEGLAND	ENDOCRINESYSTEM
ESTROGEN	GONADOTROPIC	HORMONE
HYPOTHALMUS	INSULIN	OVARY
PANCREAS	PARATHYROID	PITUITARY
TESTES	TESTOSTERONE	THROXINE

Reproduction and Development

```
N O I T A L U V O Y V M E D E
A O O C Z S S H T T E J M R T
V T I K H U U R O N Q N B O O
F C F T T R E R S V O S R C G
N U M E A B O T E I A U Y L Y
S D F W U Z R M T T O R O A Z
P I V P A U I A O O U J Y C M
E V W B A T U L M S V Y G I E
R O Q L A R N V I N O O E L N
M C R Y T I A E G T I M N I O
T L A S H P X C C F R O E B P
A M N I O T I C S A C E N M A
X E S C R O T U M X L J F U U
M R R N S E T S E T G P D P S
E B U T N A I P O L L A F F E
```

AMNION	AMNIOTICSAC	CHROMOSOME
EMBRYO	FALLOPIANTUBE	FERTILIZATION
FETUS	MENOPAUSE	MENSTRUAL
MENSTRUATION	OVARY	OVIDUCT
OVULATION	PLACENTA	PUBERTY
SCROTUM	SPERM	TESTES
UMBILICALCORD	UTERUS	ZYGOTE

Diseases, Defense and Disorders

```
L T J A P M D C I E T I N X A
E O A L N E S R A U O O H Q N
N X N L A T K U M N I H U J O
I I T E C S I O N T C I O Y R
C N I R I Y R B C A R E V U E
C D B G N S F E O E T B R I F
A V I Y Q E F W D D W E T S R
V H O X V N Y I P G Y F T V E
B T T Q I U M A L I G N A N T
Y T I N U M M I L A R U T A N
H V C I U M B E N I G N N B I
I G J N T I V P K S X T S O H
U N I N E G O N I C R A C L W
U T I N F L A M M A T I O N G
Y N E G I T N A R C K X R H L
```

ALLERGY   ANTIBIOTIC   ANTIBODY
ANTIGEN   AQUIREDIMMUNITY   BENIGN
CANCER   CARCINOGEN   HOST
IMMUNESYSTEM   INFECTION   INFLAMMATION
INTERFERON   MALIGNANT   NATURALIMMUNITY
TETANUS   TOXIN   TUMOR

Drugs, Alcohol and Tobacco

```
L  C  H  I  R  E  C  E  L  E  V  S  L  U  F  B  L  T  C  H
H  A  E  E  S  N  N  I  C  Q  E  T  A  C  A  R  B  O  N  X
Z  J  C  U  R  I  Z  N  J  T  P  J  C  O  X  F  K  J  B  X
F  O  B  I  T  O  A  S  A  T  R  D  I  D  V  J  A  I  H  O
E  A  P  O  S  R  I  R  L  N  E  S  G  O  G  E  Y  H  V  P
O  C  C  L  E  Y  U  N  M  A  S  L  O  Z  T  E  R  U  V  V
H  I  N  L  P  T  H  C  Z  S  C  Q  L  H  R  B  X  H  N  E
N  N  O  E  I  U  R  P  C  S  R  T  O  W  Z  X  G  T  E  Z
S  T  E  B  D  A  L  L  M  E  I  N  H  W  S  C  K  H  N  T
K  I  R  G  C  N  P  V  I  R  P  A  C  H  G  D  D  E  Z  W
N  A  S  K  O  J  E  L  N  P  T  L  Y  W  V  C  C  L  X  M
B  Q  Q  O  H  N  V  P  K  E  I  U  S  I  Y  O  X  T  P  D
L  D  S  V  H  A  I  H  E  D  O  M  P  T  A  U  T  R  C  V
D  M  Z  N  J  R  D  C  D  D  N  I  U  H  L  N  Q  K  G  K
W  U  A  R  B  O  R  C  U  P  E  T  Z  D  C  T  V  R  U  J
D  H  T  E  C  Q  I  I  R  L  B  S  N  R  O  E  P  D  R  W
E  D  I  X  O  N  O  M  C  R  L  N  R  A  H  R  U  T  D  Z
H  G  R  N  E  H  S  D  F  U  M  A  H  W  O  T  J  M  J  O
E  N  B  E  G  Y  M  J  T  H  W  Y  H  A  L  Y  H  I  R  Z
W  Z  R  A  P  F  X  Z  J  L  U  V  G  L  O  M  Z  P  N  W
```

ABUSE	DEPRESSANT	PRESCRIPTION
ALCOHOL	DRUG	PSYCHOLOGICAL
BARBITURATES	HALLUCINOGEN	STIMULANT
CARBON	HEROIN	THE
CIRRHOSIS	MONOXIDE	TOLERANCE
COUNTER	NICOTINE	WITHDRAWAL
CRACK	OVER	
DEPENDENCE	PHYSICAL	

Genetics

```
P O L L I N A T I O N R E G D
B R O G E R G V E F F E P E P
E A O O E T R V K S X P Y N R
N C S B M N I D T V Q L T O L
L N N E A S E I J H H I O T Y
Z E E A S B A T Y J N C N Y L
K P D E N R I P I I J A E P X
D S C N T I O L T C N T H E C
I E F Q E T M R I B S I P F K
R D S R H M O O Z T X O K Z E
B E E E I G M L D Y Y N E R F
Y D S S E E T E L P M O C N I
H I Z N D O M I N A N T K C A
S M C R O S S V J Z O A R S V
B J I C F H R V X Q O R D C D
```

BASE	CROSS	DOMINANCE
DOMINANT	GENETICS	GENOTYPE
GREGOR	HYBRID	HYPOTHESIS
INCOMPLETE	MENDEL	NITROGEN
PHENOTYPE	POLLINATION	PROBABILITY
RECESSIVE	REPLICATION	TRAITS

Genetics and Genetic Engineering

```
E  I  G  M  E  I  O  S  I  S  Q  G  B  N  H
T  L  N  N  U  P  W  R  P  I  N  H  L  O  Y
I  X  E  B  I  K  L  U  O  I  C  I  I  I  B
A  A  B  L  R  D  R  A  R  G  N  O  N  T  R
R  S  E  X  L  E  E  E  S  K  I  K  D  A  I
T  E  M  R  B  A  E  E  E  M  C  V  N  N  D
D  V  F  R  A  N  E  D  R  C  I  A  E  I  R
R  K  E  N  I  B  Y  L  I  B  E  D  S  B  C
R  E  R  G  I  W  M  S  P  N  A  N  S  M  I
D  O  N  R  T  U  J  N  N  I  G  R  P  O  T
P  E  V  I  T  C  E  L  E  S  T  A  N  C  E
R  E  P  L  I  C  A  T  I  O  N  L  D  E  N
V  R  O  L  O  C  G  U  E  D  R  C  U  R  E
K  C  U  K  R  Z  L  T  X  N  D  X  A  M  G
Z  R  V  W  N  E  K  G  O  S  R  V  V  A  Z
```

BLINDNESS
DNA
HYBRID
MEIOSIS
PUREBREED
SELECTIVE
VIGOR

BREEDING
ENGINEERING
INBREEDING
MULTIPLEALLELE
RECOMBINATION
SEX

COLOR
GENETIC
LINKED
PLASMID
REPLICATION
TRAIT

The Theory of Evolution

```
O  G  Y  Q  L  X  V  I  E  O  N  H  E  A  L
D  M  R  O  W  B  Z  V  V  O  P  O  F  N  I
P  E  K  A  Z  A  O  W  I  C  V  M  I  A  S
C  U  R  G  D  L  L  T  T  W  G  O  L  T  S
Q  L  Q  F  U  U  A  L  C  H  D  L  F  O  O
N  M  O  T  L  T  A  X  A  R  I  O  L  M  F
W  O  I  C  P  A  N  L  O  C  A  G  A  Y  U
J  O  I  A  K  A  V  C  I  G  E  O  H  A  F
N  A  D  T  T  S  E  U  D  S  U  U  A  R  C
P  A  Q  U  C  R  Y  G  A  G  M  S  I  A  S
D  J  R  I  X  E  M  B  R  Y  O  L  O  G  Y
R  A  L  U  C  E  L  O  M  D  A  T  I  N  G
L  I  S  W  S  V  R  E  L  A  T  I  V  E  Q
E  T  A  M  I  R  P  A  S  V  B  F  O  Q  O
G  D  D  V  L  R  G  R  V  I  I  N  F  N  Y
```

ADAPTATION   ALFRED   ANATOMY
CLOCK   DATING   EMBRYOLOGY
EVOLUTION   FOSSIL   GRADUALISM
HALFLIFE   HOMOLOGOUS   MOLECULAR
NATURAL   PRIMATE   RADIOACTIVE
RECORD   RELATIVE   SELECTION
WALLACE

Energy and Living Resources

```
P H X O V N U R M P F Y A L E
R Y R T C I U U A U M L H A N
W D O Q P N E C S L T A X M E
H R P K Y L O I L E O Q I R R
A O V A O L O I R E J S E E G
J E M R A N B N T F A Q T H Y
M L T L T K A V R C K R M T T
W E M R E T W U U T N R D O W
P C I M I J S V I W F I J E V
K T P V S P E C I E S K T G F
X R E D E R E G N A D N E X J
O I T N A L P O S S H G D F E
A C X Z P L F P B B J E O F K
A K V D H I A B D I Q Q R K W
V M X Q X M Q R J Q W G G I S
```

ALTERNATIVE

EXTINCTION

HYDROELECTRIC

PLANT

ENDANGERED

FUSION

NUCLEAR

SOLAR

ENERGY

GEOTHERMAL

PETROLEUM

SPECIES

Ecosystem Relationships

```
S  J  L  D  L  U  R  P  U  R  J  M  U  T  D
C  Y  I  J  W  A  A  O  O  B  S  D  V  Z  C
N  H  M  R  Q  R  C  P  T  I  P  C  S  C  Y
O  R  I  B  A  O  Q  I  T  C  O  J  T  N  U
I  W  T  S  I  E  G  I  G  M  A  P  S  P  U
S  F  I  Q  B  O  S  P  M  O  A  F  O  J  I
S  T  N  S  K  A  S  E  L  T  L  W  H  S  I
E  B  G  C  R  Y  N  I  M  L  W  O  D  H  V
C  J  W  A  Y  S  T  T  S  F  Y  Z  C  Z  Z
C  L  P  E  A  R  O  T  A  D  E  R  P  E  I
U  W  R  L  N  O  I  T  I  T  E  P  M  O  C
S  P  I  T  Y  M  Q  Y  D  Y  C  G  P  U  J
E  S  M  U  T  U  A  L  I  S  M  P  N  G  O
M  P  I  M  G  S  O  I  K  K  J  Q  R  N  S
U  T  P  O  H  K  J  H  G  J  M  L  I  P  P
```

COMMENSALISM	COMPETITION	ECOLOGICAL
FACTOR	HOST	LIMITING
MUTUALISM	PARASITE	PARASITISM
PREDATOR	PREY	SUCCESSION
SYMBIOSIS		

Biomes

```
G  C  T  H  U  U  D  K  X  S  B  N  S  N  H
R  A  V  S  J  Y  S  I  U  B  T  O  Y  Q  E
A  A  K  D  B  L  T  O  M  Q  J  T  F  I  X
S  J  I  Y  C  S  R  U  Q  T  F  K  D  Q  X
S  H  Y  N  E  E  R  T  D  D  S  N  I  P  S
L  E  P  R  F  F  R  E  S  H  W  A  T  E  R
A  J  O  I  E  O  E  T  A  M  I  L  C  T  C
N  F  N  A  P  F  R  O  W  H  A  P  I  U  Q
D  O  A  I  D  E  S  E  R  T  C  O  Q  N  T
C  M  C  N  B  T  R  L  S  Y  O  T  G  D  W
K  A  A  I  N  A  G  I  A  T  C  Y  B  R  F
L  D  O  R  A  A  P  E  U  R  X  H  S  A  J
U  M  E  B  I  V  V  A  W  F  E  P  Y  O  H
E  S  E  N  O  N  O  A  Z  J  K  I  Y  T  Q
J  T  V  F  U  Q  E  H  S  V  A  P  G  G  U
```

BIOME	CANOPY	CLIMATE
CONIFEROUS	DESERT	FOREST
FRESHWATER	GRASSLAND	MARINE
PHYTOPLANKTO	RAINFOREST	SAVANNA
TAIGA	TROPICAL	TUNDRA

Living Things, Food and Energy

```
P  E  H  Q  C  S  S  M  E  G  Z  M  U  S  Z
O  S  E  R  O  V  I  N  R  A  C  T  M  M  O
P  O  J  E  H  H  S  E  H  C  I  N  X  M  R
U  P  H  C  A  R  E  K  K  L  B  K  I  O  G
L  M  E  U  B  U  H  T  U  B  W  V  T  J  Y
A  O  R  D  I  X  T  L  E  M  O  C  A  G  V
T  C  B  O  T  D  N  U  E  R  A  U  O  H  Y
I  E  I  R  A  H  Y  T  E  F  T  L  I  D  G
O  D  V  P  T  B  S  W  C  O  O  R  A  L  G
N  P  O  J  D  Y  O  I  T  C  I  H  O  X  V
H  D  R  E  S  Y  T  R  E  T  X  X  W  P  V
H  A  E  O  F  O  O  D  C  H  A  I  N  S  H
W  R  C  R  I  P  H  R  E  M  U  S  N  O  C
Y  E  B  B  H  U  P  F  O  O  D  W  E  B  L
J  J  A  E  N  V  I  R  O  N  M  E  N  T  B
```

ABIOTICFACTOR   AUTOTROPH   CARNIVORE
CONSUMER   DECOMPOSE   ECOLOGY
ECOSYSTEM   ENVIRONMENT   FOODCHAIN
FOODWEB   HABITAT   HERBIVORE
HETERTROPH   NICHE   OMIVORE
PHOTOSYNTHESIS   POPULATION   PRODUCER

Soil, Air and Water

```
J F W K T P T V E F V L N E J
J C A D A H E G R E U I R R R
I G E R E E N S U Z A E N O E
R N N R M I R J T R Z O L S N
H E M I P I R B A I I N R I E
L A T P C T N Y R T C O A O W
L I O A E A W G E D W I Q N A
B R S T W A R L P I N T D R B
C C S S T D P R M C Z U S E L
G A H E O N R E A Y L M T E
W E R U D F A U T T T L O T C
D O H A Z A R D O U S O G I M
N O I S R E V N I R Q P X L N
C O N T O U R X Z W G D N I W
S T R I P V L I O S P O T I C
```

ACID
CROPPING
FARMING
GROUNDWATER
LITTER
RAIN
SMOG
TERRACING
TOXIC
WIND

BREAK
DEPLETION
FOSSIL
HAZARDOUS
PESTICIDE
RENEWABLE
STRIP
THERMAL
WASTE

CONTOUR
EROSION
FUEL
INVERSION
POLLUTION
SHED
TEMPERATUR
TOPSOIL
WATER

WORD SEARCH SOLUTIONS

Science

```
D + + + + + + + + + M M + +
A + + + + + + + + I + I + +
T + + + + + + C + + C + +
A + + + + + R + + R + +
+ S I S E H T O P Y H + O + +
+ E T + + B + G + + + O + +
+ + L + H I + O + + + R + +
+ + + B O E L + + L + + G + +
+ + + L A O O + + O + + A + +
+ + O + C I + R + R + + N + +
+ G + E + + R + Y T + + I + +
Y + + + + + A + N + + S + +
+ B O T A N Y + V O + + M + +
+ + + + + + + + + C + + + + +
+ + + + + + + + + + + + + +
```

Scientific Measurements

```
C + + T + + K + + + + + N +
E + + N + + + I + + + + O + D
L + E E M M I L L I L I T E R
S + + M + A + + + O S + N + +
I + + E U + S + + R G S + + +
U + + R + L + S E + I R + + +
S + + U + + O V + T + + A + +
+ + + S + + N V Y + + + + M +
+ + + A + O Y T I V A R G + +
+ + + E C I + H + + + + + +
+ + + M T + + G M E T E R + +
+ + + N + + + I + + + + + +
+ + E + + + + E + + + + + + +
+ C + + + + + W + + + + + +
+ + + + + + + + + + + + + +
```

75

Characteristics of Living Things

```
+ M A S E X U A L + N + + + N
+ E + + + + + + S O S + + O
+ T + W + + + U + I + + + I
+ A D + A + + L + T T + + D T
+ B + E + R U + A + I + I + S
+ O + + D M M T + + T G + + E
+ L + + I O S B + + E + + + G
+ I + T + O O + L S P + + + N
+ S S + E + + L T O M + + + I
+ M + M + + + I B + O A + + +
+ + O + + + O + + D C D T + +
+ H + + + N + + + + L + E O +
R E S P I R A T I O N + D M
N O I T C U D O R P E R C + +
N O I T E R C X E + + + + + +
```

Chemistry of Living Things

```
+ + + + R F + N + + + + + R
D + + E + + O O + + + E + F A
+ N T + + + + R + + L + E + G
N A U + + + + T M U + R N + U
W O + O + + + C C U R + E + S
+ + T + P + + E + U L + U + +
+ + + O + M L L M + + A T O M
+ + + + R O O E + + + + R + +
+ + + + M P + C + + + + O + +
+ + + + + E L E M E N T N + +
+ + + + + + + + + + + + + + +
+ + + R E T T A M + + + + + +
+ + + + + + + + + + + + + + +
+ + + + + + + + + + + + + + +
+ + + + + + + + + + + + + + +
```

Cell Structure and Function

```
C + M + + + + + + T + L + O
A H + S + + + + E S + + L + R
S I R + A + + L A P R S Y + G
U + R O + L L L R I + U H M A
E + + D M E P O B + + L P E N
L + + N O T O + + + O O T I
C + + A R O S + T + + E R A S
U + G O P O H O + Y + L O B M
N R L L M + + C M + C C L O V
O H A E + + + O E + U H L A
C S R A E L C U N C + N C I C
M + H O M E O S T A T I S S U
E S O L U L L E C + + I + M O
L Y S O S O M E + + + + M + L
P R O T E I N + + + + + + E
```

Cell Process

```
N + O + + + + + + C + N + E +
+ O E S A H P A T E M O + S +
M + I + M + + E + N + I + A +
+ E + T + O S + O T + T + H +
E + T + C A S I + R + A + P +
+ S + A H U S I C I + R + A +
+ + A P B S D H S O + I + N +
+ + O H U O R O + L + P + A +
+ R + F P O L + R E + S + + +
P + F + M O + I + P + E + + +
+ I + A + + L + S + E R + + +
D + T + + + + E + M + R + + +
+ I + + + + M I T O S I S + +
N I N T E R P H A S E + + + +
+ + + + + + + + + + + + + + +
```

Classification of Living Things

```
F + + + + + N + A + S + + + + A + T + R
A + + + + + O + + U + U + + R + A + A M
M + + + + + I + + + T + E I + X + L + O
I + + + + + T + + + + O S A O + U + S D
L + + + + M A + + + + T T N L + + P G
Y + + + U + C + + + O + O R L N + + E N
+ + + L + + I + + T + M + E O + I + C I
+ + Y + + + F + L + Y + C + + P + L I K
+ H O + + H I E + + + I + + + + H + E +
P + R + + P S + + + N M O N E R A N S +
+ + D + + O S + + U + + + + + + + + + +
+ + E + + R A + + + + + + + + + + + + +
+ + R + + T L R A L U L L E C I T L U M
+ + + + + O C + + + + + + + + + + + + +
+ + + + + R + + + + + + + + + + + + + +
+ + + + + E + + + + + + + + + + + + + +
+ + + + + T + + + + + + + + + + + + + +
N I T A L E + + + + + + + + + + + + + +
+ + + + + H + + + + + + + + + + + + + +
+ + + + + + + + + + + + + + + + + + + +
```

Viruses and Bacteria

```
+ + E + + + + F + R + S + N +
+ C + G + + L + E + + N O H +
+ + Y + A A + T + + + I + O +
+ V + T G H C + + + T X P S +
+ + I E O A P + + A + O A T +
+ + L R B P + O Z + + T R + +
+ L + + U + L I I + + + A + +
A + + + + S R A + + + + S + +
+ + + + + U + + S + + + I + +
+ + + + E + + + + M + + T + +
+ A N T I B I O T I C S E + +
+ + S + + + + + + + + + S + +
+ A + + + + + + + + + + + + +
P + A I R E T C A B + + + + +
+ + + + + + + + C O C C I + +
```

Protozoan

```
S U E L C U N O R C I M N S C
T + N + E + + E + M + A + U O
+ S + A + U T + U + R + + E N
+ + A + O I G I + E + N P L J
+ C + L S Z C L F + A + L C U
T + I A P E O I E O + E A U G
+ O R L M O C T Z N T T S N A
S A P A I I R O O E A A M O T
P P R S M A R O L R + L O R I
+ A O A E O T L L + P L D C O
P + R R P Y U E + A + E I A N
+ O + S E G E + + + B G U M +
F D O P O D U E S P + A M + +
E N I D O C R A S + + L + + +
E L C I L L E P + + + F + + +
```

Algae and Fungi

```
+ + + + + + + + + + + B + P F Y + +
+ + + + + + + + + + U I H + U E R + +
+ + + + + B + F + + L N O T K N A L P +
+ + R A L U L L E C I T L U M G S E + +
+ + + + + D + R + R O + U + A I T + + +
+ + + + + D + + A S M G M + G A S + + +
+ + + + + I M + Y L + E I + L + + + + +
+ + + + + N O N + + U + N L A + + + + +
+ + + + + G T + + + + C E T L + + + + +
+ + H Y P H A E + + R G S + A + + + + +
+ + + S E T I S A R A P C A + T + + + +
+ + + S + + D D + L L + E + V + I + M +
T + I + + + L F K U B N + + N + O + +
+ S L L Y H P O R O L H C + + + O + N +
+ + A + + + N M + A L A E + + R + N +
+ + + F + I + + D + E + T + H + + + +
+ + + + D + + D + + C + + S + + + + +
+ + + + + L E + + + I U + + + + + + +
+ + + + E R O P S + N M + + + + + + + +
+ + + + S + + H + + U + + + + + + + +
```

Vascular Plants

```
G D N O R F + + + E + N + G
E S E M O Z I H R S + O + Y E
R R + M + + S + O + I + M P L
M + E + U I + C + T + N + H C
I + + P M I U + A + O I + L I
N + + P R L B Z + S + R + O T
A + L + G O I M P Y + O + E U
T E + + L D E A + R S + M C
I + + + I + R U + C + A + + +
O + + T + M + + C + + + V + +
N X R + S T O M A T A + + O +
+ E Y + + + + + + + I + + + +
F + + G + + + + + + + O + + +
G I B B E R E L L I N S N + +
+ + + + + N + + + + + + + + +
```

Ferns

```
+ + + E + + + + + + + + + + +
+ + + T S + + + + + + + + + +
+ + + Y + O + + + D + + + + +
+ + S H D + R S N + + + + + +
+ + + P N O P I + + + + + + +
+ A + O O O W + + + + + + + +
+ + S T R R + + + + R + + + +
+ + S E F A O + + G H + + + +
+ + + M X L + P + + A I + + +
+ + + A + U + S H + M Z + + +
+ + + G + C A + N Y E O + + +
+ + + + + S + L + R T M + + +
+ + + + + A + + + E E + + + +
+ + + + + V + + + S F + + + +
+ + + + + + + + + + + + + + +
```

Seed Plants

```
+ + + X Y C + + + R + D + S +
+ + + Y G U + + I + R + I + +
C + + L N T + N + A M S A + C
S H + E O I G + U R E N + + H
I + L M P C + G E H G + + S L
M + + O S L + P T I + + T + O
R + + + R E S N O + + O + + R
E D E E S O Y S C + M + + + O
D T R A N S P I R A T I O N P
I + + Y O E + H T M M + + + L
P + G T R + + A Y W E B + + A
E + O M T U B E R L O O I + S
  H P A L I S A D E L O L U T
P S U O E C A B R E H + D H M
N E G Y X O + + + + + + + Y P
```

Gymnosperms and Angiosperms

```
P R Y R A V O + + C N A L S N
S O E + + + F O O + N I T O
+ T L W + + R T I + + G T I I
+ + Y L O U Y T + + + I S G T
+ + + L I L A + + L + O I M A
+ + + T E N F + A + + S P A Z
+ + + D I + A T + S E P A L I
+ + O M + + E T + + + E + + L
+ N R + + P + + I + + R + + I
+ E + S R E I F N O C M + N T
G S T A M E N + + + N E E + R
G Y M N O S P E R M S L + + E
+ + + + + + + + + + L U + + F
+ + + + + + + + + O + V + + +
+ + + + + + + + P + + O + + +
```

Invertebrates

```
N E M A T O C Y S T + N + + E
+ + E X O S K E L E T O N H L
+ + T R I C H I N O S I S O T
P L A T Y H E L M I N T H S N
E A M D O P O R H T R A + T A
G P O R I F E R A N + Z + + M
A U R E G E N E R A T I O N A
L P P + S E D E P I L L I M R
F P H E R O M O N E S I + + A
O P O R C S E D E P I T N E C
M + S N A E C A T S U R C + H
A + I E T A R B E T R E V N I
C H S I F Y L L E J + F + + N
P O R E S D I L E N N A + + D
C O R A L S N E M A T O D E S
```

More about Invertebrates

```
E C N O T E L E K S O X E F J
T R M I L L I P E D E S L + E
A U D R A Z Z I G + + A S + L
R S + + S E D E P I T N E C L
B T + + P T L + + Y + + G + Y
E A + O + + C A H + + + N + F
T C R E G E N E R A T I O N I
R E R A D U L A S V + + P + S
E A + + + M + + + N A + S S H
V N N O I T A Z I L I T R E F
N S E N O M O R E H P + + T +
I + T A R T H R O P O D + A +
+ H T R I C H I N O S I S E +
S E D O T A M E N + + + + + +
S T A R F I S H + + + + + + +
```

Coldblooded Vertebrates

```
T L O M V E R T E B R A E + +
A M P H I B I A N + + + + + +
E X T E R N A L + + + + + + N
+ + + + R E P T I L E S O D
N O T E L E K S O D N E I C E
+ B L + R + G M + E + T S A D
+ A + A + E O N T + A + E R O
+ C + + N N D A A N + + L T O
+ K + + E R D N R F + + O I L
+ B + V + R E E A + + + P L B
+ O + + O + B T + M + + D A M
+ N + H + I + + N + A + A G R
+ E + + H H S I F I + L T E A
D E D O O L B D L O C + A + W
R E D D A L B M I W S + + S +
```

Warm-blooded Vertebrates

```
+ T + E + B R + L + W H E M E
A + E B R U I A + A + E R A N
+ T A R O O I R R + M D O M I
C T N T R P V M D I + G V M N
S A N E U I B I G S + E I A A
+ O S S C L T R N + + H B R C
C + R R O A A O + R + O R Y W
+ A + O I T L + R + A G E S H
M + D + E A + P + Y + C H R A
+ E O L L I D A M R A + + E L
D E T A B U C N I + + + + V E
I N C I S O R A B B I T + A +
+ + E T A M I R P + + + + E +
+ + + + + + + + + + + + B +
N W O D + + M O N O T R E M E
```

The Skeletal and Muscular Systems

```
O + C C A S R U M E F L N L C
C R + A L L M + + + + I A A O
A W G + L A L O + + + G I T N
R O T A + C V E O + + A S E N
D R + N N E I I T T + M R L E
I R + + I S G U C A H E E E C
A A + + + O Y A M L P N V K T
C M + + + + J S L I E T A S I
O S S I F I C A T I O N H + V
T + + + + + + H + E T + + + E
E + V E R T E B R A M R + + V
N + P H A L A N G E S + A + R
D + + + I S U R E M U H + C E
O + + A M U E T S O I R E P N
N E L B I D N A M U S C L E +
```

The Digestive System

```
+ D + S B P R M C + G + L C L
P + I I A O E H O A + A + A A
+ T L G S E E P L L C E N R R
N E Y I E M R L S I A I + B G
A O C A I S B C N I M R + O E
V N I C L L T A N A N O S H D
I I A T A I H I T A + L I Y I
L L L D P C N I O + P A T D G
A + D L E R V + + N + C T R E
S E A M I N O A C I D S O A S
R S I S L A T S I R E P L T T
T N E I R T U N B C + + G E I
H C A M O T S + E A + + I + V
+ C A N I N E F + + + + P + E
+ + + M I N E R A L I V E R +
```

The Circulatory System

```
A T R I U M + + + + + + Y + A
T R A N S F U S I O N R + T +
V N + H + + + + + O + H + P
L E O + E L C I R T N E V A C
O T I I + M S F A + R + C + A
R P E N S E O L I O + E + + P
E L + L P N U G S B M + Y C I
T A H T E C E C L A R R + I L
S S U P R T L T K O E I E N L
E M + I M E A E R T B V N O A
L A C + R Y R L R E L I + R R
O + + O + + L A P A P + N H Y
H + S A T R O A V + + Y + C +
C I + + + + + W H I T E H + +
S R E T E M O N O M G Y H P S
```

The Respiratory and Excretory Systems

```
+ N + + S + + + + + E + A L E
V A O + I + E + + X + L U X E
O E Y I T T + X H + V N C M L
C R E + T + R A C E G R G N U
A U N L O A L A O E E A O + S
L + D I L E R L C T R R + + P
C U I V G + U I O H H T + + A
O R K E I S + R P P E + I + C
R E + R P + Y A E S + A + O +
D T + + E S I N + T E + + + N
+ H + + Y D E P I D E R M I S
+ R + S + + + N O S T R I L +
+ A T S U H C N O R B N I K S
R E D D A L B Y R A N I R U +
M L A R Y N X E L A H N I + +
```

The Nervous System

```
+ L + D + M M + R + + + S D A
N + A + E U E E + + + E X R E
+ E + R R N C T C + N F E O L
M + R D E E D I S S + F L C H
+ O R V P H M R O Y + E F L C
+ A T T O O P R I + S C E A O
E + O O N U Y I L T + T R N C
+ R + O R N S + R A E O + I P
+ + T A E N R O C E R R S P U
+ U S U L U M I T S P T Y S P
A + R C E R E B R U M + N + I
+ O M U L L E B E R E C A E L
N + + + M E D U L L A X P + C
S N E L R E T I N A O + S + +
+ + + + + + + + N + + E + +
```

The Endocrine System

```
E + + + P A R A T H Y R O I D
+ N T E S T E S + + + + + + S
+ E D G + P E + + + + + + + U
E N D O C R I N E S Y S T E M
+ O + N C + O T I + + + + + L
+ M + A + R + V U X N + E + A
+ R A D + + I + A I O S + + H
+ O D O + + + N L R T R + + T
+ H R T + + + U E R Y A H + O
+ + E R + + S + O G + + R T P
+ + N O + N + G + + L + + Y Y
+ + A P I + E + + + + A + + H
+ + L I + N + S A E R C N A P
+ + S C + + + + + + + + D +
E N O R E T S O T S E T + + +
```

86

Reproduction and Development

```
N O I T A L U V O Y + M E D E
+ O + C + S S + T + E + M R T
+ T I + H U U R O N + N B O O
+ C + T T R E R S V O + R C G
+ U + E A B O T E I A + Y L Y
S D F + U Z R M T T + R O A Z
P I + P A U I A O + U + Y C M
E V + + A T U L M S + + + I E
R O + L + R N + I N O + + L N
M + + + T + + E + T I M + I O
+ + + S + + + + C + R O E B P
A M N I O T I C S A C E N M A
+ E S C R O T U M + L + F U U
M + + + S E T S E T + P + + S
E B U T N A I P O L L A F + E
```

Disease, Defense and Disorders

```
+ T + A + M + C + + T + N + A
E O A L N E S + A U + O + Q N
N X N L + T + U M N I + U + O
I I T E + S I O N T C I + + R
C N I R + Y R B C A R E + + E
C + B G + S + E O E T + R + F
A + I Y + E F + D D + E + + R
V + O + + N + I + + Y + T + E
+ + T + I U M A L I G N A N T
Y T I N U M M I L A R U T A N
+ + C + U M B E N I G N + + I
+ + + N + I + + + + + T S O H
+ + I N E G O N I C R A C + +
+ T I N F L A M M A T I O N +
Y N E G I T N A + + + + + + +
```

87

Drugs, Alcohol, and Tobacco

```
N O I T A L U V O Y + M E D E
+ O + C + S S + T + E + M R T
+ T I + H U U R O N + N B O O
+ C + T T R E R S V O + R C G
+ U + E A B O T E I A + Y L Y
S D F + U Z R M T T + R O A Z
P I + P A U I A O + U + Y C M
E V + + A T U L M S + + + I E
R O + L + R N + I N O + + L N
M + + + T + + E + T I M + I O
+ + + S + + + + C + R O E B P
A M N I O T I C S A C E N M A
+ E S C R O T U M + L + F U U
M + + + S E T S E T + P + + S
E B U T N A I P O L L A F + E
```

Genetic

```
P O L L I N A T I O N R E G +
B R O G E R G + E + + E P E +
E A O + E + + V + S + P Y N +
+ C S B + N I + T + + L T O +
L + N E A S E I + H + I O T +
+ E + A S B A T Y + N C N Y +
+ + D E N R I P I I + A E P +
D + C N T I O L T C + T H E +
I E + + E T M R I + S I P + +
R + + + H M O O + T + O + + +
B + + E + G + + D + Y N + + +
Y + S + E E T E L P M O C N I
H I + N D O M I N A N T + + +
S + C R O S S + + + + + + + +
+ + + + + + + + + + + + + + +
```

Genetics and Genetic Engineering

```
E I G M E I O S I S + G B N H
T L N N + P + R P + N + L O Y
I + E B I + L U O I + I I I B
A + + L R D R A R G N + N T R
R S E X L E E E S K I + D A I
T + + B A E E E M + V N N D
+ + + R + N E D R + I + E I +
+ + E + I + + L I B + D S B C
+ E + G + + + + P N + + S M I
D + N + + + + + I G + + O T
+ E V I T C E L E S T A + C E
R E P L I C A T I O N L + E N
+ R O L O C + + + D + + U R E
+ + + + + + + + + + + + M G
+ + + + + + + + + + + + + +
```

The Theory of Evolution

```
+ G + + + + + E + N H E A L
D + R + W + + V V O + O F N I
+ E + A + A O + I + + M I A S
C + R + D L L T T + + O L T S
+ L + F U U A L C + D L F O O
N + O T L T A + A R + O L M F
+ O I C P A N L O C + G A Y +
+ O I A K A + C I + E O H + +
N + D T T + E + D S + U + + +
+ A + U C R + + A + M S + + +
+ + R + + E M B R Y O L O G Y
R A L U C E L O M D A T I N G
L + + + + R E L A T I V E +
E T A M I R P + S + + + + + +
+ + + + + + + + + + + + + + +
```

Energy and Living Resources

```
+ H + + + N + R M + F + A L E
+ Y + + + + U U A U + L + A N
+ D + + + N E C S L T + + M E
+ R + + + L O I L E O + + R R
+ O + + O + O I R E + S + E G
+ E + R + N + N T + A + + H Y
+ L T + + + A + + C + R + T +
+ E + + + T + + + + N + O +
P C + + I + + + + + + I + E +
+ T + V S P E C I E S + T G +
+ R E D E R E G N A D N E X +
+ I T N A L P + + + + + + E
+ C + + + + + + + + + + + +
+ + + + + + + + + + + + + +
+ + + + + + + + + + + + + +
```

Ecosystem Relationships

```
S + L + L + R P + + + M + + +
+ Y I + + A A O + + S + + + +
N + M + + R C + T I + C + + +
O + I B A + + I T C O + T + +
I + T S I + + I G M A + S + +
S + I + + O S + M O + F O + +
S T N + + A S E + + L + H + +
E + G + R + N I + + + O + + +
C + + A Y S + + S + + + C + +
C + P E A R O T A D E R P E +
U + R L N O I T I T E P M O C
S P I + + + + + + + + + + + +
+ S M U T U A L I S M + + + +
M + + + + + + + + + + + + + +
+ + + + + + + + + + + + + + +
```

Biomes

```
G + + + + + + + S + N + + +
R + + + + + + + U + + O + + +
A A + + + + T O + + + T + + +
S + I + + S R + + + K + + +
S + Y N E + T + + + N + + +
L + P R F F R E S H W A T E R
A + O I + O E T A M I L C T +
N F N + P + R + + + + P + U +
D O A I D E S E R T + O + N +
C M C N B + + + S + + T + D +
+ A A I N A G I A T + Y + R +
L + O R + A + + + + + H + A +
+ M + + I + V + + + + P + + +
E + + + N + A + + + + + + +
+ + + + + E + S + + + + +
```

Living Things, Food and Energy

```
P E + + + S + + + + + + + +
O S E R O V I N R A C + + + O
P O + E H H S E H C I N + M R
U P H C A + E + + + + I O +
L M E U B + H T + + + V T + Y
A O R D I + T + E M O C A G +
T C B O T + N + E R A U O + +
I E I R A + Y T E F T L + + +
O D V P T + S + C O O R + + +
N + O + + Y O I T C + + O + +
+ + R + S + T R E + + + + P +
+ + E O F O O D C H A I N + H
+ + C + I P H R E M U S N O C
+ E + B H + P F O O D W E B +
+ + A E N V I R O N M E N T +
```

Soil, Air and Water

```
+  F  +  K  +  P  T  +  E  F  +  +  N  E  +
+  +  A  +  A  H  E  G  R  +  U  I  +  R  R
+  G  +  R  E  E  N  S  U  +  A  E  N  O  E
R  +  N  R  M  I  R  +  T  R  +  O  L  S  N
+  E  M  I  P  I  +  B  A  I  I  N  +  I  E
L  A  T  P  C  +  N  +  R  T  C  O  +  O  W
L  I  O  A  E  A  W  G  E  D  +  I  +  N  A
+  R  S  T  W  A  R  L  P  I  +  T  D  R  B
C  +  S  S  T  D  P  R  M  C  +  U  S  E  L
+  A  H  E  O  E  N  +  E  A  +  L  M  T  E
W  E  R  +  D  F  +  U  T  T  T  L  O  T  +
D  +  H  A  Z  A  R  D  O  U  S  O  G  I  +
N  O  I  S  R  E  V  N  I  R  +  P  X  L  +
C  O  N  T  O  U  R  +  +  +  G  D  N  I  W
S  T  R  I  P  +  L  I  O  S  P  O  T  +  C
```

CROSSWORD SOLUTIONS

Science

Scientific Measurement

Characteristics and Needs of Living Things

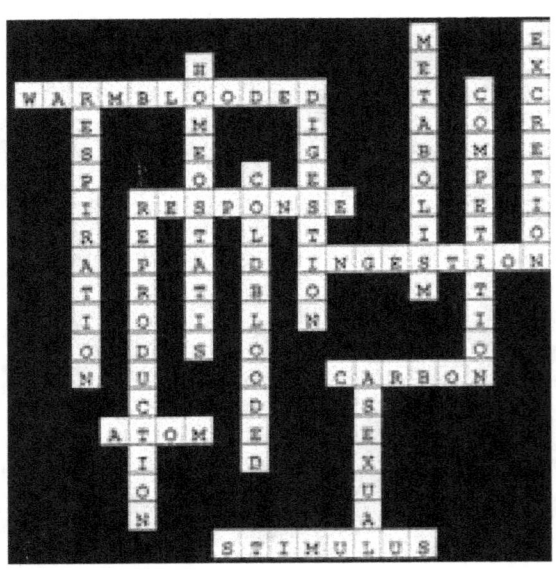

Cell Structure and Function

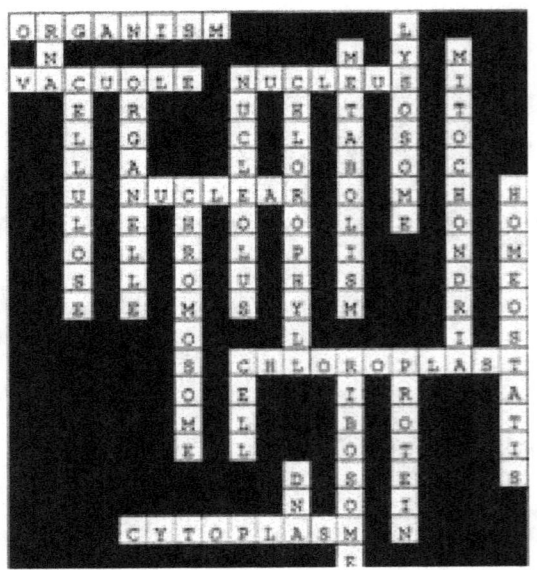

Chemistry of Living Things

Cell Processes

Classification of Living Things

Viruses and Bacteria

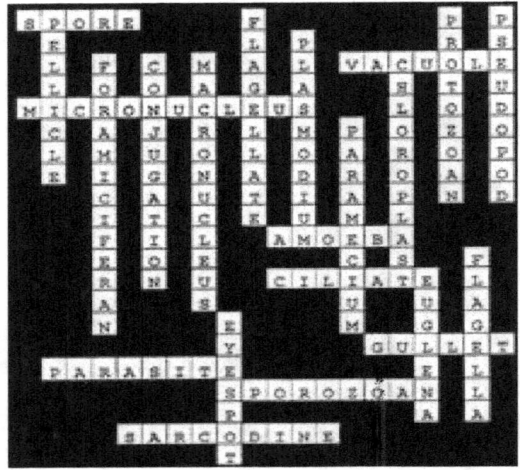

Protozoans

Vascular Plants

Ferns

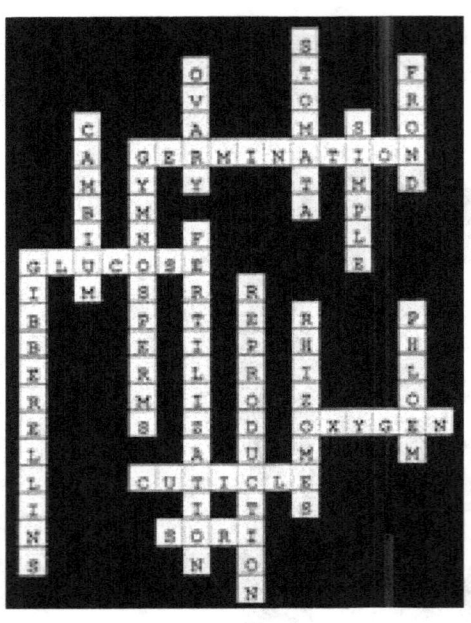

Algae and Fungi

Seeds and Plants

Gymnosperms and Angiosperms

Invertebrates

More About Invertebrates

Cold-Blooded Vertebrates

Warm-Blooded Vertebrates

Skeletal and Muscular Systems

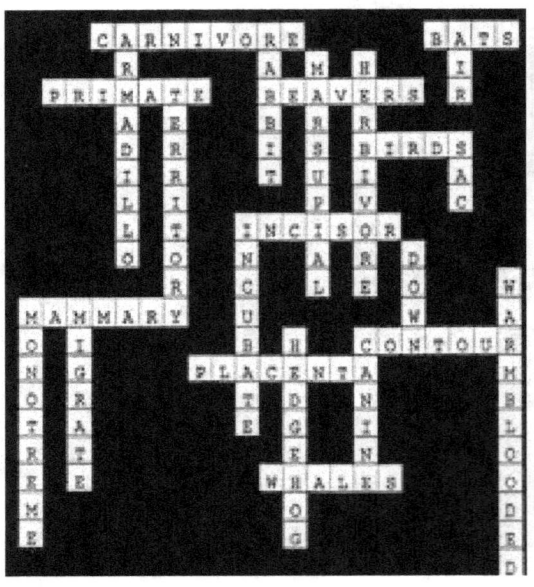

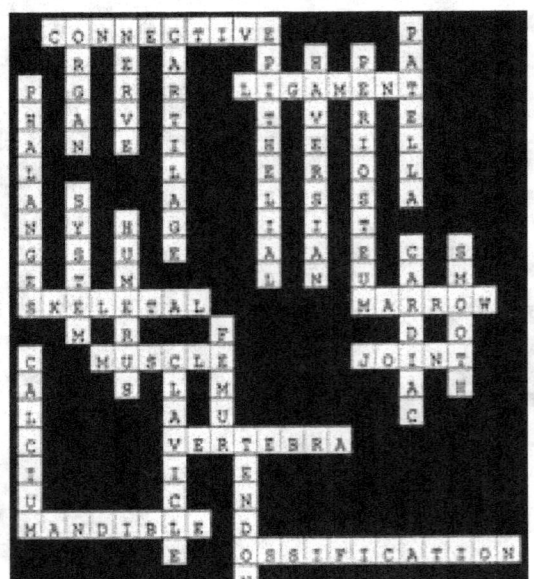

Nervous System

Circulatory System

Respiratory and Excretory Systems

Endocrine System

Reproductive System

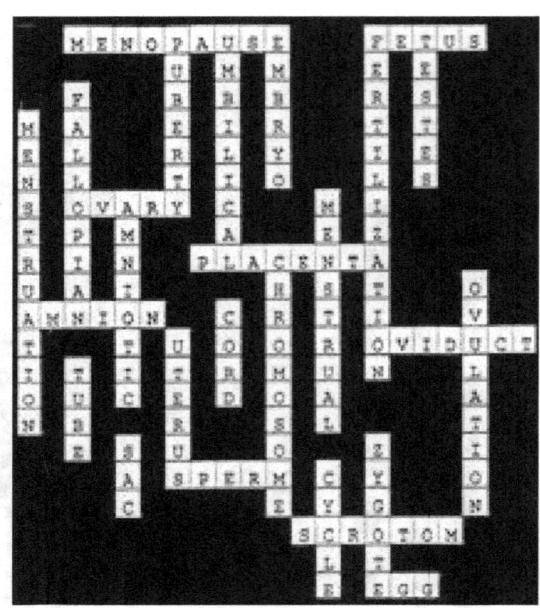

Diseases, Defenses, and Disorders

Drugs, Alcohol, and Tobacco

Theory of Evolution

Genetics and Genetic Engineering

Energy and Living Resources

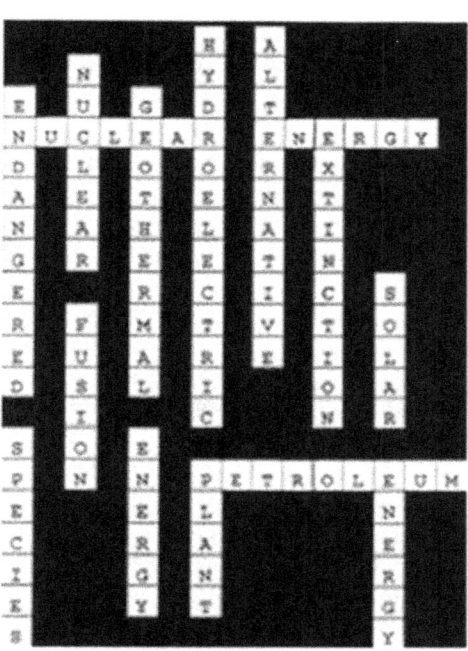

Genetics

Living Things, Food and Energy

Biomes

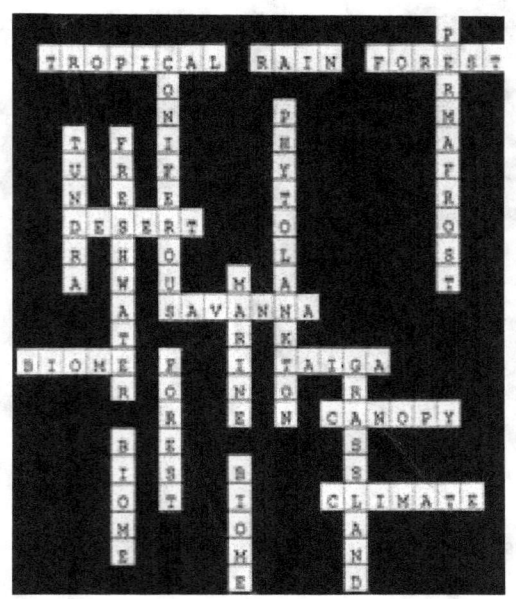

Soil, Air, and Water Conservation

Reproduction and Development

Word List

- sperm
- chromosome
- ovary
- oviduct
- ovulation
- placenta
- menopause
- amnion
- egg
- testes
- Fallopian tube
- uterus
- embryo
- fetus
- zygote
- umbilical cord
- fertilization
- scrotum
- menstrual
- amniotic sac
- puberty
- menstruation

Across

1 physical change in females after which menstruation and ovulation stop
5 developing baby from the 8th week until birth
9 endocrine gland that produces female hormones
12 structure through which developing mammals receive food and oxygen while in the mother .
15 clear membrane that forms a fluid-filled sac around the embryo in the uterus
18 fallopian tube; the tube through which an egg travels from the ovary
22 male sex cell
24 external sac in males that contains the testes
25 female sex cell

Down

2 beginning of adolescence
3-16 structure that connects the embryo to its mother and transports oxygen and wastes   (2 words)
4 newly formed organism that is the product of fertilization
5 joining of the egg and the sperm
6 male sex glands
7-19 tube through which an egg travels from the ovary (2 words)
8 process in which the blood and tissue from the thickened lining of the uterus pass out of a female's body through the vagina
10-20 fluid-filled sac that cushions and protects the developing baby in the uterus (2 words)
ll-23 monthly cycle of change that occurs in the female reproductive system (2 words}
13 rod shaped cell structure found in the center of cells carry long pieces of DNA
14 Process in which an egg is released from the   ovary into the Fallopian tube
I7 pear-Shaped structure in which the early development of a baby takes place
21 fertil1zed egg